2K15 Dec

The Life Etheric with Carol Croft

goto: w3 ETHERIC WARRIORS.COM

ALSO: goto: CRYSTAL INSIGHTS.NET

ALSO: DB's SITE w3 thevine.net/cbswork Pg 112t

w3 WORLDWITHOUTPARASITES.COM Pg 145n

ALSO: LIFE QUEST FORMULAS.COM

Sylph Photos: w3 RYANMCGINITY.COM Pg 133t

see ORGONISEAFRICA: w3 Pg 133b

USE OF CHEMBUSTER FOR HEALING Py 303b AND WORLD W/O PARASITES.COM AND UNCURABLE.COM Py 304m

Pg 218b - w3 RYANMcginty.com WEBSITE OF SYLPH PHOTOS

Pg 304m - w3 INCURABLE.COM w3 LIFEQUESTFORMULAS.COM DETOX HEALING INFO

By Don Croft

TRAFFORD PUBLISHING

27t - ZAPPERS CURES

ZAPPER: 15Hz Ref: WORLD WITHOUT PARASITES.COM

42b - SASQUATCH CALLS WEBSITE

1ST READING 2K16 FEB 19

2ND READING 2K17 AUG 01

NOTE: COPYRIGHT 2K6

Note for Librarians: A cataloguing record for this book is available from Library and Archives Canada at www.collectionscanada.ca/amicus/index-e.html
ISBN 1-4120-9565-4

Offices in Canada, USA, Ireland and UK

Book sales for North America and international:
Trafford Publishing, 6E–2333 Government St.,
Victoria, BC V8T 4P4 CANADA
phone 250 383 6864 (toll-free 1 888 232 4444)
fax 250 383 6804; email to orders@trafford.com
Book sales in Europe:
Trafford Publishing (UK) Limited, 9 Park End Street, 2nd Floor
Oxford, UK OX1 1HH UNITED KINGDOM
phone +44 (0)1865 722 113 (local rate 0845 230 9601)
facsimile +44 (0)1865 722 868; info.uk@trafford.com
Order online at:
trafford.com/06-1320

10 9 8 7 6 5 4 3 2 1

"Blessed are the meek, for they shall inherit the earth."

Jesus Christ

"...The satanic secrets and deeds done in the gloom of night shall be laid bare and manifest before the peoples of the world."

Baha'u'llah (1817-1893)

P277m – AT&T a CIA FRONT
277m – (=) ANTENNAS ΔT P – ΔT ANTENNAS
301b – RADIATION NEUTRALIZATION!
302t

313b CURING – AN ACT OF AGGRESSION
(HEALING – AN ACT OF LOVE

314t } CHRONIC DISEASES – LYME, FIBROMYALGIA, EBOLA,
315b
316t } MADCOW, HEP C,
340t – TESLA COILS!!!...

354m – ZAPPERS DESTROY PRIONS!

119m → ORGONITE C.B.

327m – HBP

132t – SORCHA FAAL
170t – (=) HHG
171t (=) PORK = (+) ORGONE ENERGY The LIFE FORCE
173b – TOM BEARDEN
109b
313b – CURING – AN ACT OF AGGRESSION
VS
HEALING ACT OF LOVE

141b – (=) SONY

HBP

Acknowledgments

Dr Wilhelm Reich wasn't adequately appreciated when he was alive, and by the time he died, in 1957 he was essentially erased from public record or misrepresented by cynical officials in government, academia and science. But since this global healing network started in the spring of 2001, his life work has steadily gained serious, international interest again.

Indications are good that this kindly, gifted and productive genius will finally get the recognition he earned during his lifetime, as the world's premier researcher and inventor, in the study and uses of subtle energy in the fields of psychiatry, medicine, biology, physics, chemistry and meteorology. All of us are now riding on his coat tails.

In the US, it's against federal law, (sic) to even have orgone-based healing devices, let alone make and distribute them, as scores of people are now doing. Who are the real lawbreakers in this case? Don't you agree that it's time to stop allowing unelected, national bureaucrats and unlawful secret police to crucify healers and pioneers like Dr. Reich?

Judy Lubulwa, an erudite Buganda princess who lives in Nairobi and may be setting a new, global standard for the transformation of arid wasteland back into vibrant, productive farmland, through the distribution of simple, inexpensive orgonite devices, has put her heart into editing this book, in a thorough and timely way. She suggested that the following caveat ought to be included somewhere, so I figure that this page is the best place:

WARNING: This book is not for the faint hearted, nor for those who want others to tell them what to do. Some of the information in this book is highly unusual. Please do not read any further if you want to stay in your comfort zone and are happy with what you read, see and hear in the media. Read on if you are open to the possibility that things could be different.

Table of Contents

Chapter One

Self-Empowerment (Really)

At the end, my highest distinction may be that I was generally noted for having had a good marriage with a telepath. A lesser distinction, is my talent for stating the obvious, such as, 'The meek are inheriting the earth.'

In five years, a global, grassroots network has grown up around the simple activity of making and tossing cheap, simple orgonite devices into the environment and then observing some phenomenal but consistent results. Carol and I started this empowering movement, but we don't own or control it, of course. We're a couple of beta personalities, not glory hounds or chest-pounders, and we simply wanted to share some empowering information via the internet.

I'm offering this book as encouragement for you to empower yourself by making and distributing simple orgonite and for observing the astonishing environmental and social transformations that will follow. Thousands of people have been doing this around the world in relative obscurity until now. This simple process has turned out to be our most significant discovery, so far, in the six years of our very productive, (and never boring), partnership and marriage.

Simple instructions for making and distributing inexpensive orgonite devices are available as a free download from www.ethericwarriors.com. Real-time instruction for learning to toss energy effectively from the heart, for healing or for correcting injustice, is also available, if you contact anyone who posts on that site. In that case, you'll be directed to a competent psychic, who will likely charge you a nominal fee to watch remotely, as you learn the process and to then give you pointers until you get it right. Energy tossing is simple to learn and anyone can do it effectively. If the psychics didn't charge a little money for this, they'd burn out before long, and we all want them to remain in the game, of course. What you'll learn will be worth many times more than what you'll be paying to learn it.

I wonder if you can conceive what it's like to be married to a telepathic woman. I've enjoyed the privilege of our close association since June 2000, and I suppose that if I were a secretive person, it would seem more like a punishment than a love life. According to the Latin root of the word, 'privilege,' private laws were written for or against a private person in ancient Rome. The word has come to have an entirely positive connotation but, in fact, with every blessing there's added accountability, which is often burdensome. 'Blessing' and 'curse' can sometimes be seen as relative terms in that context but, for the record, I just feel mightily blessed by being married to Carol.

I chose, early on, to look at the brighter side of this phenomenon and I've grown to love this remarkable, selfless woman more and more, as time and experience move along. Carol, the more gifted one, has been pretty much the silent partner in this marriage so I've been beating her drum on the internet whenever possible; now I'm able to honor her with a book. I don't dwell on anyone's imperfections, including my own, because I have faith that by rather dwelling on our better attributes, those will shine bright enough to draw attention away from our shortcomings, which then seem to shrink from neglect. As with any other spiritual principle, this one is reflected by scientific and social dynamics.

The contribution by my wife and other reputable psychics and energy sensitives to the expansion of this grassroots orgonite network, has been a priceless benefit and entirely essential. They're able to see or accurately sense subtle energy dynamics, so they give us immediate confirmations when we're on the right track with energy experiments and new inventions. Having a good psychic as a research partner is a 'next level' approach beyond having to rely on mere physical instruments. In the case of subtle energy research and development, even the best instruments can only detect the electromagnetic expressions of the process, which are always secondary to the underlying, more powerful etheric dynamics.

As intelligence gatherers, trained and gifted psychics can't be beat, which is why the quasi-government agencies such as the CIA, MI5/6 and their equivalents in the East rely so heavily on their own psychics these days. The trained, gifted psychics who work for 'the other side', are a literal army who greatly outnumber the more reputable, (surviving), psychics who won't cross over.

Remote viewing is relatively passé, of course, and so is publicly trotted out by disinformants as 'cutting edge' intelligence gathering protocols. Learning to remote view might be helpful for those of us who aren't particularly psi-gifted, but to say that a remote viewer can hold a candle to a disciplined psychic is pretty absurd. Like dowsing, remote viewing can add materially to the empirical database, though.

Fortunately for us, working with orgonite and perhaps especially the selfless service aspect of this network's adventure, tends to unlock latent psi ability when a person has that undeveloped talent. One psychic who has personal integrity and a conscience can stand against any number of opponents in the etheric realm, because predators simply lack substance and are only able to exploit personal weaknesses in their opponents. So, actively opposing these predators is merely a good exercise in character development, when one or two of them manage to cause us pain. You absolutely can neutralize them, as many of us around the world are now doing. This is the other 'public service' we've been performing in addition to healing the environment and boosting humanity's awareness.

Each of us can stand successfully against the combined forces of the predatory world order, of course, but they aim their best shots, (including frequent poison attacks, neutralized by zappers, mostly), at our psychics, because the other side apparently believe that we can't do this work without the psychics' help. We can do it without the psychics, actually, but it's just not as much fun or enlightening without their participation.

We're each born with certain latent talents and capacities. I wasn't born with the gift of high psychism, so I don't angst over the lack. Carol wasn't born with my passionate curiosity and she doesn't feel a lack in that respect. I suppose that if I were constantly flooded with psi data I'd be less curious. When most of us look at someone we see a body; when a psychic looks he or she sees the energy around the body, also the hitchhikers and they also see ghosts and other ethereal life forms, otherwise. At night, while driving she has a hard time distinguishing between living people and ghosts, for instance, so when a ghost occasionally shows up in her headlights its pretty challenging, and she often has to quickly decide whether to just drive on or slam on her brakes. She makes that determination, usually, by looking at their clothing.

If you're still a thrall of the What To Think Network, you might assume that what I'm telling you is debatable because for you, in that case, nothing is ever certain. If you have a chance to be close to a psychic, day by day, you'll soon learn that these things are simply facts of life.

Everyone has flashes of psi awareness through our intuition, of course, but that's something that occurs in times of need, mostly. We need that a lot, when we've broken free from the What To Think Network.

Rare folks like Carol can look into the ethers, read animals', ghosts', aliens' and people's minds, and they can pop out of their bodies any time they want. Fortunately for the rest of us, perhaps, most psychics are so overwhelmed by psi input most of the time, that they don't want to look into the heads of other people. Carol only does it when she can't avoid it and when we're under attack.

Some psychics smoke cigarettes just to slow down the constant avalanche of data, in fact. Sometimes when she's driving long distances she keeps a heavy object in her lap to remind her to stay in her body, though she's capable of driving on the highway while out of body.

Any informal group of energy tossers within the orgonite network, which includes one or two psychics, can get a lot done, directly interfering with government/corporate predators' mass murder plots and otherwise weakening persistent, old but hidden, parasitic hierarchies, including the disgusting espionage/assassination agencies I referred to. Having two or more reputable psychics on board during these informal sessions most often enables consensus, which puts their psi impressions in the category of empirical evidence. It also generates synergy, which is a marvel to behold and participate in.

The meek are inheriting the earth, as long promised, and anyone can participate in this grassroots, spontaneous and unorganized effort. Imagine how it will be when we've decided to stop enabling the cancer of centralized power, and can keep our local and regional governments in check by quietly neutralizing predators, who have managed to infiltrate them. Tossing energy this way is a healing process. The receiver of the energy can either heal or be neutralized, depending on his/her orientation at the moment. Predators, especially cold-blooded murderers, generally have sacrificed their volition in favor of gaining temporary advantages over others, so they're easy meat for meek energy tossers like us.

Ken Adachi, of www.educate-yourself.org had graciously posted 90 of my journal reports, between June 2001 and November 2005, in which I wrote about our, (Carol's, my and a few companions'), subtle-energy-related adventures and projects during that time. The book you're now reading will include some focus on our current work with the dolphins and whales, which began in earnest in December 2005, for Carol.

The 90 reports are now available in Word format for no charge on www.ethericwarriors.com, which is the network's sole surviving public record. If you want them on your site, too, please contact the Etheric Warriors webmaster.

In April 2001, Carol and I were astrally visited by couple of bottlenose dolphins from the research facility across the Overseas Highway, (US Route 1), from our R/V campsite one evening in the Florida Keys. That was the evening following the night that we tossed some simple orgonite into a couple of those dolphin pens on Grassy Key, Florida, in a daring, (read: trespassing), midnight sortie, from the Gulf side of that narrow island in our dinghy. We risked being caught, because it was unlikely that we'd be able to toss the orgonite into those pens during the day, due to the watchfulness of the staff.

We were both shown some astonishing things in our waking state, by those two dolphins then, and because we were both wide-awake, we were able to compare notes in real time.

That experience was a reciprocal gift from the dolphins to Carol and I and it's taken us this long to get back here—we moved to Jupiter, Florida in September 2005. We intend to finish some business in the Bahamas and Yucatan, before we leave here, and that might be the subject of another book.

We first came to South Florida in December 2000, because of some insights we had received from 'The Operators' on and near Mt Shasta, during the fall equinox, 2000, involving specific locations in the Bahamas, the Texas coast and Yucatan. D Bradley invented that term to express the fact that benevolent 'someones', are apparently watching over this network and guiding our steps through our own intuitive processes.

That compelling, equinox glimpse of a larger, underlying reality—the earth's energy grid and its relationship to humanity's awareness and development—provided the impetus for the subsequent, seemingly accidental unfolding of this global, grassroots, orgonite network, during the years that have followed. You're probably reading this on account of what's being done by that expanding, revolutionary movement, even if your first glimpse of us might have been on account of a character assault by a newsreader, or clever disinformant on national TV, which I've characterized as The What To Think Network for the purpose of this book. I've borrowed that term from a "MR SHOW" episode. Most of what we're presenting came from other sources—all we deserve credit for is combining it into a consistently empowering process and sharing the info.

Dr Wilhelm Reich is the one who deserves the most credit, because he made the universe's energy matrix a subject of productive, scientific inquiry, when he was alive. His work is the foundation of this network.

When we arrived in Florida, I'd built a seaworthy, efficient motor skiff, but it was a little too small to take Carol and I across the Gulf Stream comfortably, so I went to the Bahamas alone in it, from Miami. I got caught up in a HAARP storm and also became disoriented by the erratic energy of the big vortex east of Bimini, so I failed to complete my task. Carol and I returned to South Andros Island five months later, where I'd left the boat, but the seas remained too rough for us to reach our objective.

This time, we bought a twenty-three foot long Zodiac Pro 650, which is practically unsinkable and has a 600-mile range. When the seas flatten out more, pretty soon, we'll be able to make it to Bimini from Miami, in about two hours, and from Key West to Yucatan, in about six hours—probably after this book is published. I banged that Zodiac up pretty good, gifting in the sea over the winter, and it's in the shop now, being fixed under warranty. I thought it was supposed to be able to take that punishment, but 'live and learn,' eh? Carol's a better driver than I am, and I'm not ashamed to admit it.

Carol's dolphin friend, five years before, showed her some underwater, accessible, Atlantean ruins, south of Andros Island, and also showed her where it is on the chart. I believe it's the same location described by someone who visited there in the early 1970s and retrieved some artifacts. My dolphin visitor mostly just showed me colors and gave me a lot of wild love and energy—what a rush! The cool part is that we were both wide awake and discussing it. This backs up the old, pre-patriarchy understanding that women are mainly mental, and men are mainly emotional. Successfully turning that around, through forced ideologies, thousands of years ago in Europe and Asia, was one of the nails in humanity's temporary coffin of cultural schizophrenia, I think. Fortunately for us, that happened only recently and ineffectively, among the Black and Red people.
We also intend to drop some orgonite on the huge underwater

base that the US and British Navies maintain, in the Tongue of the Ocean, 3,000 feet down. It's marked with a beacon buoy on the charts. No kidding. There are several support facilities for that underwater base on South Andros Island, where the people get very nervous if you stop to chat. I found that out directly, when I stopped at one of them in my skiff, in December 2001. Carol and I are both quite eager to finish our original mission. The trip to Yucatan from Key West is over a couple hundred miles of open water, and this is the area where HAARP boosts its HAARPicanes, so we'll have some fun hurting that agenda, too.

Richard Sauter, a research journalist who uncovered the reality of the new underground base network in the world, asked me a few years ago for evidence of that underwater base, but of course I don't have direct evidence, so he didn't report it, though he remains interested and open-minded.

We're still all trying to discern what, precisely, we're doing in this invigorating movement, but the unfolding discovery process is much of the fun, of course, and lots of new participants, perhaps including you, are bringing their own puzzle pieces to add to the big picture these days. The latest development during our chat sessions is that very old and large elementals are spontaneously getting involved with us, in our efforts to take down the major occult/corporate/government/military predators. Jeff McKinley triggered that last fall, when Carol and I arrived in Florida. He'd been inspired to build a special device, but didn't know what to do with it until Carol saw a huge elemental on the Yucatan coast, attached to it.

Jeff immediately asked that elemental to push down on the eye of Hurricane Wilma, which was passing by Yucatan at the time, and the eye was seen to widen significantly after that, in radar imagery. We got the other folks in our informal group involved and tracked several more events like that, timed with our efforts. By the time it reached shore at Naples, Florida, it was no longer a hurricane. It came at us in a straight line, across the state, and built up momentum due to HAARP transmissions from around the regions, but when it reached 'ground zero,' where Jeff and we live, it had immediately diminished to Level Three tropical storm, thanks to Jeff's previous tower-busting efforts in the area. That storm caused very little damage here, but south of the eye in Ft Lauderdale, the winds were much stronger, due to still-unbusted HAARP facilities in that area.

Eric Carlson, Jeff and I went throughout the area along the path of Wilma, and disabled all of the HAARP arrays and death transmitters, a few months later. As you probably know, natural hurricanes lose momentum fast over land, rather than gain momentum, and they never travel in a straight line.

When we moved here, the What To Think Network was already predicting progressively more damaging and even longer 'hurricane seasons', for the Gulf region; now they're saying that the Gulf will probably be spared hurricanes, for the foreseeable future. How funny is that? You, too, can prevent violent storms where you live, if you'll go out and disable all of the HAARP arrays and death towers with cheap, homemade orgonite devices. It really is that simple, as many are discovering. Some of us are even reversing deserts now.

Carol says that the budding partnership between us little guys and those big elementals, in taking down global tyranny and healing our wonderful planet, is simply possible because humanity are ready for more direct interaction and partnership now. We tentatively take this to mean that the tide has turned irrevocably against this old patriarchic hierarchy, too. Lots of our associates are getting that feeling by now.
The new science of orgone dynamics, whose foundation was

substantially laid by Dr Wilhelm Reich in the 1930s, is partly also an art, because orgone, (an arbitrary term introduced by Dr Reich to describe the matrix of the seen and unseen worlds), is alive, creative and seems to have awareness, so interacts, often playfully, with the researcher.

Genuine research physicists have known for generations, that there's actually no such thing as objectivity, of course, but a practical level of scientific method is still necessary. We're essentially rational beings, after all, and history teaches us that one must never accept anything that conflicts with one's rational faculty. The new synthesis of spirituality and science breaks down the former 'rational' paradigm, which is glorified 'denial,' and crass materialism. Theosophy's pitiful 'irrationalism,' which is thinly disguised, blind belief in infantile dogma, is also a casualty of the process, thank God. We need to honor our rational faculties and our own heart promptings, or we spin out into delusion, paranoia, conformity and hopelessness.

Being married to a skilled and experienced, high psychic has had a bleed-through effect on my own preternatural perceptions, from time to time. I get enough confirmations this way, to have faith in what Carol and the few other reputable psychics are telling us in real time, during our more arcane experiences. Here's a rule of thumb that we've found useful: if any psychic, no matter how powerful, has no control over his/her personal life, the data is always suspect, but if a psychic has demonstrated consistency and humility, the data is always usable.

One's reputation is built on one's personal integrity, over time, and we put a lot of stock in reputations in this network, as anyone ought to.

Carol's particularly good at popping out of her body and back in safely, and she's on an intimate basis with some of The Operators, so she's usually able to answer my spontaneous requests for psi intel, during the day. In cases where the answer is hidden in the details or is beyond her comprehension, an Operator will often give her specific guidance, when we're in need. We all get that through our intuition, as I mentioned, but sometimes it's handy to just ask and receive through a reputable psychic. Often, we're obviously not supposed to have specific answers, and in that case, we have to trust The Operators' judgment and be detached.

It's important to establish that no psychics' statements are authoritative on their own, of course. I'm offering this advice in the context of personal inquiry. Nobody can have the authority to say what's true or not, and this is one of the features of the emerged new paradigm. Clergy and arbitrary authority in general, are as obsolete as chastity belts.

Discernment is a struggle that none of us escape, because we generally have to earn what we get, and that's in order. Psychics have no more or less discernment ability or 'spirituality' than non-psychics and, as I said, the majority of psychics work for the bad guys these days and are locked into dogmatic, 'irrationalist', (Theosophy is the tired, ersatz global religion of the occult/corporate order), ideologies. Even if they were to come over to our side, they would probably just muddy the waters for years, until they can get up to speed.

Nobody, much, simply 'switches sides,' because working for the other side, at least for Europoids and Asians, involves many years of deep, trauma-based programming, that causes clinical schizophrenia, which needs healing over time. Many, many millions of people have been intentionally programmed this way, as small children, usually without the parents' being aware of it. I recommend getting information on this from people like Al Bielek and some of the writers who are published by Franklin Press; most of the people who write about Montauk, Tavistock and the CIA's Monarch Program are disinformants, otherwise.

The old families, whom these disinformants work for, originated 'irrationalism,' which is one of the older mind control protocols, after all; based loosely on an amalgam of Hindu, Buddhist, Vril and a few dualistic, (read: schizophrenic), ideologies. I consider essential Hinduism and Buddhism to be valid, creative forces in the world, and I think it's a crying shame that the British took some of their offerings out of context, in the middle 1800s, and twisted them to serve the ideological needs of a parasitic world order, as a false religion, adding the ancient, veiled, but satanic Vril practices, and silly dualism, ('the world is evil'), into the mix.

Documented conspiracy information is consistent and makes sense. I recommend that you seek it out. The What To Think Network constantly broadcasts non-conspiracy theory, which is the institutional denial of human history's deeper dynamics. Best-selling books like "THE DA VINCI CODE", which may well be evidence of the occult/corporate world order's 'strategic retreat', in the face of rising human awareness, can perhaps be seen as evidence that the What To Think Network is losing ground fast, by now, even though the books are essentially puff pieces and even recruiting tools for irrationalism and secret societies.

My instincts are pretty good by now, because there have been times when trusting them was the only option, so I committed to following my instincts without question about ten years ago, after a personal trauma. I went through a similar period as a youth, but my marriage to a predator pretty much put my personal development on hold, for 22 years after that. I was blessed with four children during that time, so I'm not complaining.

Once, in my youth, (1969-70), I was subjected to the unfriendly, meddlesome attention of Vril sorcerers, in Germany. They got cranky and vindictive, when I refused to join them, and that was my first introduction to the dynamics of the occult/corporate world order. I was in the US Army. Another time, not long after that, I was harassed by voodoo practitioners, in Louisiana and it happened again in the Caribbean.

I endured a hurricane in a small, open sailboat in the Gulf of Mexico, for a few days, in the autumn of 1995. A couple of firewalks were comparatively easy later on, near Seattle, in late 1999. Firewalks are an awful lot of fun, actually, and if you get a chance to stroll barefoot, on a bed of glowing embers, you should! It's a good way to escape the What To Think Network's oppressive mental shackles, for a bit, and to get a broader view of reality in a short time. The only thing that can stop a PJ person from remaining asleep, short of Divine intervention, is a committed desire to wake up. Not even a firewalk or a visit to a space ship can do that.

Our heartfelt, (higher), instincts are what directly connect us to The Operators, as I said. Psi ability alone, can't achieve that. Only the psychics, whose hearts are open to beneficial influences, will feel that connection. The rest will remain caged in their minds, and will enthusiastically regard triggered squirts of endorphins, via programmed, 'irrationalist' stimuli, and lockstep new-age conformity, for spirituality and certitude. How pitiful is that, really? We can do better now.

In the years before Carol and I got together I had committed to following the promptings of my instincts, without question, as a sort of enquiry into the nature of common intuition, and it was quite a lonely, but productive, three year process. Carol was my reward for that commitment, and she's moved with me into a much broader commitment, which is global healing and service with orgonite and the attendant, lawful neutralization of predatory hierarchies. We've shared that with all comers, since some of us began developing the protocols, because all of this is just something we're ready for and is needed. By now, the protocols have expanded and involve the contributions of many committed and talented folks, of course, all of whom I'll tell you about. It's not complete, because as our awareness continues to expand, new opportunities for healing are discovered and, not least, this isn't an organization.

Carol passionately avoids confrontation, and prefers to focus on the healing and awareness aspect of the movement. Her healing inventions reflect that. However, when she's personally attacked by humanity's enemies, she takes them down efficiently and aggressively. This was the case with a county prosecutor and Sheriff in Northern Idaho, who tried to frame and imprison her daughter last year, in a federally-sponsored effort to diminish Carol's viability within the network. Our group efforts have turned the tables on many similar plots, over the years, which were aimed at the more effective members of the network, often doctors in that case, since they depend on licenses for their livelihood.

I tend to be more concerned about exposing and removing obstacles to planetary healing: mainly the movers and shakers within the occult/corporate world order's hierarchies. I also gift a lot, of course. I don't think that one can be very effective in this network, unless one pursues both paths at the same time, although it's not essential to work in groups. Most people who do this work are alone by choice, in fact.

When I go after predators it's usually as part of a group effort, though I'm pretty handy, as are the others, in fending off psychic predators in my daily life and initiating some reasonable reprisals on their chain of command.

This simple activity will come to be seen as commonplace, in coming days, and it will spell the end of institutional exploitation forever. It's a more spontaneous and refined version of 'The Minutemen' of the American Revolution, I think, and is in keeping with the emerging global paradigm. This revolution is essentially non-violent and ultimately more effective than shooting predators, although I suppose some of them may face capital punishment for their crimes, in due course and within the law.

The most grievous obstacle to anyone recognizing the true nature of the widespread problem of exploitation is the hypnotic, enervating What To Think Network, of course. In the West this obstacle is best represented by television, now that people have mostly grown beyond our dependence on clergy and other arbitrary authorities. I think that in the East, self-policing programming is mainly ancient and cultural. Note, for instance, how closely Mao Tse Tung tailored his manipulation of the masses, on ancient Taoist manipulation protocols, thinly disguised as Marxism, which is infantile.

It's well known that Taoism is a technical approach to reality; it's not a religion and it has no ideology. It's simply information, which is neither good nor bad. Lao Tse provides the 'high road' example of the application of Toaism, in the "TAO TE CHING", which I think ought to be studied by schoolchildren, throughout the planet. At any rate, we'll need to come to terms with China, because it's likely that they've just taken over 'civilization's' initiative from the Europeans and Americans. The growing number of African gifters will soon balance out the more negative aspects of China's growing influence, I believe, and after secret police surveillance and violent oppression in China is reduced to a level that finally allows the dissemination of information, orgonite will spread like wildfire there, too.

We see Russia as China's reluctant but obedient lapdog by

now, though they, too, obviously wish to dominate the world militarily. The US is no longer a force to be reckoned with, as can be seen by our drastically diminished military and our China-based physical economy. China apparently, (foolishly), assumed the fake debt of the Federal Reserve Corporation, last fall, in exchange for that London-based corporation's old gold, which was the US Treasury before 1935. We have been conquered, but a few of the more recalcitrant PJ folks are still waving flags. My solace is in knowing that if a few states simply secede, then China's domination of our economy, through corporate manipulation, will evaporate. It's really a global economy, after all, and gold has no real value: the real assets are people, agriculture and raw material. The bad guys have always acted on that knowledge, and it's time that we do the same and stop supporting them.

Carol never lost her psi ability because she was never seduced away from it by popular paradigms; she's a very private person who keeps her own counsel, and she never takes orders. Most of the rest of us have to slog and hack our way through the veils, facades and tangled webs, which hide basic spiritual truths from us, but she and a few other natural psychics in the West, have managed to circumvent most of that, by simply keeping their mouths shut, as children. In her case, an abusive, alcoholic mother induced her to remain silent, for her own psychological survival. She didn't meet other psychics and she wasn't exposed to their lockstep irrationalism, until she was in her late twenties, so her worldview was pretty solid by then.

Her mom, a member of the Blackfoot tribe in Montana, who probably had been psychic as a child, hammered Carol whenever she expressed her psi impressions, so she kept them to herself. Her dad was quietly supportive, during her childhood and later, often bragged about her ability to his friends. His grandfather was a famous, (notorious), character on the Montana frontier. Her mother's grandmother was part Gypsy, from Austria. Often, psi ability is handed down through the generations, mainly through females.

When particularly talented psychics are children, they don't distinguish between the living and the dead, so it was problematic for Carol to talk about her experiences, especially with departed people and pets. Our own dog, six months old, was murdered with poison and a heart attack, a few weeks ago. It hit us pretty hard, but our observations afterward, showed me some interesting things about the progress of animals' spirits beyond the physical realm, and that's helped me deal with the pain of loss. In that case, Carol mentioned that the dog's spirit lingered here disoriented, for a day or so, before moving into the light, right after which she apparently returned in spirit, happy, with several other dogs. Both of us have had the sense that she's watching out for us, now, and the peculiar physical evidence is the appearance, each day, of marks from her metal collar buckle, on the toilet seats and lids. She had started drinking out of the toilets, in February, when we left the bathroom doors open. That's when she got tall enough to reach over the seat and drink the water, and we often were annoyed by the marks after that, but not any more.

I was put on the Homeland Security Abomination's official enemy list of 50,000, (well, they see all breathing people as their official enemies—let's face it—but they only list that many), right before the dog was murdered, so a little more help from the unseen realm is most welcome, at the moment.

Some professional, often charismatic disinformants on the internet and in the other media, have been trying to portray Carol and I as dangerous, 'agents,' or kooks, probably in order to distract attention from the empowering implications of our discovery: orgonite's raw potential to heal the environment and uplift the human spirit. Happy people progress spiritually and in every other way, after all, and orgonite tends to make people happy. We were already a lesser threat ,on account of our zapper business, because any zapper quickly, cheaply and easily cures cancer, AIDS, heart disease, arthritis, herpes, mental illness, etcetera. When the Feds started harassing American zapper makers, we went on the road in August 2000, and we never even got an email from those criminals. The War on Healers has been waged by the US Government, ever since that institution was handed over to the London-based corporation, in the 1930s. Dr Reich, himself, was a casualty in that one-sided war. Well, before now, it's been one-sided, of course.

If I were seeking popularity, I'd do my best to play down the more arcane part of our personal experiences, but that would make me pretty boring, and even a little dishonest. My wife and I take a lot of pleasure in exploring realms, which the What To Think Network wants you to believe are imaginary or perilous. Reporting our personal observations doubles the fun, at least for me. Since I'm the more public partner, I can take the brunt of the savagery that gets directed at Carol, from time to time. 'Trouble' is usually just another word for 'fun' for me, fortunately, and the occasional personal betrayals haven't cut me deeply. She has been hurt by a few personal betrayals, within the network, though.

Naturally, some of what we encounter is just too weird to write down, at least for now, but it's essential for everyone to get free of the non-conspiracy theories, which keep them from seeing how the world has been systematically exploited for centuries, and even millennia. The Old and New Testaments are open records of human conspiracies against God's Messengers, and the Baroque period in Italy has provided us with a wealth of literature on how this exploitation has been carried out, through the ages. The Borgia popes' brothel and satanic rituals within the Vatican, is a more contemporary example of some of the older Bible reports, for instance.

Finally understanding and appreciating real human history, is an integral step on the path to genuine economic and political freedom. Discernment gets to be fun, once you get past the raw, initial paranoia that goes along with recognizing how thoroughly we have all been scammed by occult/corporate parasites. The 'shellshock' stage only lasts a little while, if you're a fairly balanced, sober person. If not, you're likely stay in the etheric asylum until you get free of that psychological addiction.

There is no political freedom without economic freedom. Usurious families have steadily gained economic and political power since the days of Babylon, until they lately attempted to bring the curtain down, worldwide, on the very last of our birthrights: free speech and breathing. Russians and Chinese voluntarily gave up all their birthrights generations ago, after they opted to accept the political protocols of the London School of Economics' false god, (communism), and voluntarily disarm. Well, not so voluntarily, for scores of millions of them, of course—they were systematically taken down by famine, mobs and regular pogroms. That was a blow-by-blow repetition of the fake revolution in eighteenth-century France, in case you hadn't noticed.

I think that if this parasitic, global hierarchy hadn't been sucker punched by the internet in the nineties, we might have never gotten the chance to heal the planet with orgonite, because we'd all have been destroyed by their biological weaponry, summarily decapitated as 'dissidents', in those well-documented, assembly-line execution facilities, or confined to work camps across the landscape. The internet has been essential, until now, for disseminating this powerful information and for growing the network. Right before that, short-wave radio programs kept people informed, but that was eventually subverted and stopped.

These execution and mass-internment facilities, (new and fully staffed at the time: 1994), were shown on US national TV, shortly after the feds blew up the Murrah Federal Building, in Oklahoma City—remember? The internet spread across the world, not long after that, and the old world order soon lost its momentum and initiative. They've failed to get it back and I think it's our job to ensure that they won't get it back. I think we're also obligated to help this parasitic world order fall faster, since every day they're in control of the planet's resources, people are suffering and dying un-necessarily.

It's the titillating aspect of my more arcane accounts, combined with a non-intrusive, beta personality and my frequent assurance that I only ever give subjective accounts, which make this challenging material seem digestible, to the more balanced readers. For genuine, heartfelt activists who have functionally mastered their own anger and have sought ways to get the most bang for their buck, in healing society and the environment, this is all usable information, and they hit the ground running with it. Those few committed folks are our potential coworkers and friends and most of us never get past the astonishment that there are other genuine people in this world, after all, who have social conscience. We're generally programmed from birth to be cynics, after all, or to chase after irrational goals, as can be seen with the millions of energetic poseurs who bow the knee to Rockefeller-funded 'environmental' organizations, but fail to make the world better, and in fact, simply contribute to the fascist/communist agenda of centralized power.

It's not an exclusive demographic, of course, because most folks will prefer not to exercise intellectual integrity or to serve others effectively. But anyone can do so, and this little segment of society, throughout the ages, are the ones who always initiate and sustain human progress, on behalf of everyone else.

I'm appealing directly to that little demographic now, so I don't expect to sell millions of books, especially since this one is not a titillating adventure story, per se. Discernment is a bigger adventure, please consider.

Most people in the world are only concerned about their comfort and getting along with others, and there's sure no harm in that, but we like to refer to these as the Pajama People, (a title of a song by Frank Zappa,) because they studiously avoid empowering themselves or exercising discernment, for reasons more than mere survival and staying clear of discomfort. If you try to induce them to see a broader world, you'll be crucified by them, so it's best to just leave orgonite around them and love them selflessly but, otherwise, not to interfere with their sleepwalking. They're the main beneficiaries of our efforts, since they're the vast majority of humanity.

The few people who choose to exploit others might be approximately equal to the number of people, (us), who rather want to serve, heal and educate others, and to effectively remove the more grievous obstacles to dynamic unity and prosperity in our lovely world.

Until now, the few predators and parasites have manipulated the PJ folks to move against us, but they're losing control of the PJ folks, which means we're suddenly on an even playing/battle field now. That's pretty exciting, especially since we have the initiative! The bad guys are handicapped by the simple fact that harming any of us will draw the attention of the PJ folks to tyranny and exploitation. My hope and prayer is that none of us will be harmed, ever, and that global tyranny will come crashing down the way the WTC and the

Murrah Federal Building did—relatively harmlessly to those surrounding the event.

When orgonite comes into mainstream awareness, it will be the vast majority, (PJ folks), who will finish the job of ending tyranny through mere distribution, but that won't be their motivation; orgonite simply makes us all feel better and it stops the bad guys from stealing earth energy, so who wouldn't buy or make some? Orgonite showed up so suddenly and earned some favor on the net so fast, that the other side hasn't been able to scare people away from it.

As long as the vast majority, the PJ folks, are moderately content and are not worked into a frenzy by the exploiters, then the good guys, (including you?), have a chance to make this world into a relative paradise, before long. Gifting, (taking territory from the enemy), and etherically assaulting the evildoers' parasitic hierarchies, (depriving the other side of effective leadership and cohesion—they attack their own when they perceive weakness), ensures that the PJ folks will feel progressively happier and safer, which means that functional unity, peace and prosperity will arrive sooner to our specie.

In the process of distributing orgonite intelligently and systematically, the war machine of the occult/corporate order, which includes weather warfare, omnipresent death towers, biological weapon assaults and mass mind control, is being exposed and neutralized; the world will thereby run more and more smoothly. Zappers mostly remove the threat of disease, and they're simple and cheap. Affordable, low-tech free energy technology, soon to be disseminated by many inventors, will free us from dependence on the oil and nuke cartels. Economic slavery is political slavery and that's been ending.

Most of us have been scammed by ideologues into believing that some future event or even a calendar moment will mark a major turnaround or victory for the good guys over the exploiters, but in practical terms that's a process which began many years ago and has simply gained more and more momentum over the decades. To believe that this will suddenly happen in the indistinct future, is to buy into the scam that nothing good has been happening, so far, (dualism: 'the world is evil'), and that we'll somehow be rescued by—fill in the blank: Jesus, Maitreya, spacemen, photon belt, channeling, 'indigo children,' ad nauseum. As we see it, the bad guys aren't going to go out in a blaze of destruction and mayhem, defeated by supernatural agencies on our behalf; they're more likely going to continue to get worn down and exposed by orgonite, energy tossers and even the PJ folks' involuntarily-rising awareness, until it's time for the legitimate cops and soldiers around the world to just casually round them up, to stand trial for their crimes. I don't think it will necessarily happen in '2012,' either. Sheesh.

The 'Space Brothers' or Jesus aren't going to suddenly show up and solve all of our problems, especially since one of the most widespread spiritual problems in the world right now is charity addiction—voluntary enslavement—under National-Socialism and that other false god, communism. Selling 'free health care', through excessive taxation isn't working well, since millions of dying, cancerous PJ folks and their families are waking up to the stark reality that most MDs are just expensive serial killers. At least Jack Kavorkian was honest about taking people's lives for cash, and Jeffrey Dahmer 'practiced' on his victims without calling it science.

It's a lot better for us to creatively solve our own dilemmas through consultative will, mainly on a local level, where real political power should reside. That's what some in this network are doing, and The Operators have given us means to do that: instant, free global communication, orgonite and energy-tossing. The internet and cell phones make up a peerless, timely communication network, for the entire specie, and if the primary troublemakers in the world, which, (at the moment), are the occult/corporate-funded governments of China, UK, USA, Israel and Russia, will simply be exposed and dealt with by their respective populations, we'll have taken some major strides toward having a free, prosperous global commonwealth based on de-centralized political and economic power and truly representative government.

This might seem complicated to you but, really, excessively centralized government, which is the essence of exploitation by the few by intimidation, is falling under its own weight right now, as human awareness continues to expand at an accelerated rate. Even the PJ folks are waking up a bit now, though they're being driven rather than led. Case in point: the more the What To Think Network in America bleats about the 'threat of foreign terrorists,' the more obvious it is that there is no terrorist threat. Even CNN pundits lately indicate that the US Government, not swarthy Muslims, have been clearly implicated in blowing up the World Trade Center. This is simply another 'strategic retreat' ploy. Note how they are implicating the fake US president, who is too stupid to put together a simple sentence, let alone orchestrate such a complex plot.

This false government is simply losing more and more popular support, daily, and turning a few greedy, midlevel predators in the White House into sacrificial goats, on the WTC pyre, won't stop Americans from distrusting this entire treasonous regime. The Patriot Acts and volumes of other legislation are grounds for capital punishment for the authors and supporters in Congress, (the vast majority). The occult/corporate world order made an untimely bid for martial law, by terrorizing the PJ folks on 9/11, years ago, after erecting millions of death transmitters throughout the land, but they've failed to follow through and blow up more and more innocents in America, in order to direct blame at foreign Muslims and increase the PJ folks' terror, (read: eager acceptance of more and more tyranny). The Emperor is now parading naked, and I'm astonished to be the only one to publicly call this, 'treason,' in my writings, during all the years since 9/11. I noticed that most folks are programmed not to use, or even hear that word. How strange is that? I realize that by stating the obvious, in this book, my neck will be on the block as never before, but as soon as two or three other popular writers start doing so, the threat will pass, I think, and maybe I can safely live in the US again.

There are measurable, observable effects of the mega-scale healing work we're all doing, (including the lack of 'terrorist events' in the US for the past four years, thanks to our periodic, systematic, etheric disruption of the hierarchies' leadership), and thousands of reports of this were available on internet boards, until fairly recently. In English, only www.ethericwarriors.com remains, as a credible public record for people in the network, though the French and German boards are pretty solid and will also likely survive. All other orgonite reporting sites in English, have been compromised and vitiated by now; most are gone. That represents many thousands of first hand reports and observations, by people who performed healing personally, in the environment and in society.

I'm mostly reporting my subjective impressions here, so please discern everything I write, okay? If your main concern is comfort within your personal paradigm, you might not enjoy reading this book, but if you feel a need for more self-empowerment through sharper discernment, you're going to be able to use this book's information and energy to improve your life and the lives of those around you. Effective action usually follows discernment.

Most of the people who gift, but haven't yet broken free of the What To Think Network, are too intentionally blind, still, to even witness the phenomenal effects of their own gifting efforts, which is a crying shame. Getting the confirmations is requisite to self-empowerment, and to staying active in the network. One has to be genuinely empowered, after all, before one can grasp even the simplest of truths; the What To Think Network is in place to ensure that nobody comprehends anything at all.

'Preternatural' means 'beyond natural.'

Technically, much of what you'll be experiencing and observing, if you will take my advice and start distributing orgonite in your environment and tossing energy at predators, (try both at work for some instant gratification), is 'beyond normal,' but orgone dynamics seem to contradict a few of what are presented by institutionalized scientists as natural laws. A word from the dictionary is sometimes handy, when we want to appreciate a new, challenging concept. There are so many challenging, new concepts in the material we're presenting, that I'm not going to tax your credulity by also making up new words, especially since the English language is so rich.

We don't hate institutional predators, any more than a big game hunter hates animals. In fact, hatred makes an energy tosser ineffective and even vulnerable; genuine lovers of humanity actually are the best predator blasters. Weakening predators in the old, secret hierarchies is a public service; an act of love. The same energy, thrown in the same way at someone whose heart isn't set on exploitation, heals that person—sometimes dramatically—and gives him or her energy. So, even if we aim at the wrong target we're not doing any harm.

There's simply no way to lose or to harm others, when we take this initiative. The occasional predators, who fall from our efforts, are essentially committing suicide; we're simply invoking balance under universal law and this process gives us no personal advantage over anyone.

If we were to step outside the bounds of law, we'd simply be more meat for the predators, but that's not possible with these protocols. They can't hit back effectively, when we do it right, in other words.

Chapter Two

Unfettered

Carol suggests that a proper initiation into the matters of etheric healing and etheric warfare is a brief discussion of our experience with several Sasquatch, during the summer of 2001, northeast of Spokane, Washington. We arrived there two months after we had that profound psychic experience one evening with those two dolphins, in the Florida Keys—the recipients of our first 'orgonite dolphin gifts.' We had our closest encounter with a UFO as we were leaving that campground for good—it came close to hitting the R/V I was driving, in fact, and the halogen lights on the bottom of the hissing, ten meter diameter saucer gave away its earthly origin. That was fun, but I think it was designed to intimidate us.

The reason she gave for the suggestion, is that the cetaceans are mostly able to move in and out of our 3, allegedly solid, dimensions in a similar way to the Sasquatch' practice and, perhaps more significantly, one usually sees cetaceans in the wild, when one is meant to do so by them, which is also how one is able to perhaps hear, smell or see Sasquatch.

This easy ability of the more visible dolphins and whales, their wide dispersion throughout the planet's seas and their expressed willingness to teach us to be better people, has apparently caused some concern to the occult/corporate hierarchies. These two-legged predators, mainly through the US, British and Russian navies, have been systematically trying to destroy the cetaceans, with aggressive sonar and other strategies, which accounts for the occasional beachings, in past decades. Underwater death transmitters, similar to the ones that were suddenly erected throughout the world, in the fall of 2001, are routinely seen on the seabed, by the more expensive sonar devices of yachts and commercial watercraft. They're seen less clearly, even with our Zodiac's fish finder. As far as we can tell, these underwater death towers are as new and widespread as the ones on land are, and are just as easily found and disabled, we discovered. They seem to be directed at poisoning the seas with deadly orgone radiation, which is the ultimate poison.

We've only met one other person who has interacted with Sasquatch: Bob Billings, an American Indian medicine man, who has an herb shop on the highway, south of Ronan, Western Montana, on the Flathead, (Blackfoot), Reservation. Carol was told in Hawaii last year, by Joan Ocean, that someone on Maui has been interacting with Sasquatch there.

Joan Ocean and a few others began sharing dolphin contact experiences with large numbers of people, in the late 1980s and we'll probably soon find more and more abundant confirmation that the dolphins' and whales' desire for orgonite devices to disable the underwater death transmitters, is leading us to the next stage of a fruitful, interspecies partnership.

I think that before long, we'll be seeing more reports of reciprocal interactions with these phenomenal creatures, (cetaceans, Sasquatch and other sentient, but hyper dimensional species who share our planet), because the veil between our familiar dimensions and their more expanded world, is getting thinner and thinner.

It may be that we homo sapiens are the stewards of our planet, but it may also be that dolphins and other, more-etheric-than-human species, are our next teachers, as genuine global civilization continues to unfold. Maybe interplanetary cooperation will soon follow. I have no doubt that the technology is already in place, because I've seen plenty of antigravity craft, as perhaps have you.

It's just as well that I mention this in the beginning, rather than try to schmooze you into weirder considerations later on. You can get your own confirmations easy enough by distributing orgonite intelligently, after all, and your progress and earned knowledge will be from your own efforts, not from reading this or any other book.

> The root meaning of the Yiddish word, 'schmooze,' is 'to converse or chat,' and the more subtle meaning, suggests personal conversation aimed at friendly manipulation.

There are some things that can only be described with Yiddish words, and I discourage people from blaming Jews for trouble in the world. This prejudice is almost as persistent as the fear and hatred of empowered females. It might be helpful to know that the few Jews, (Zionists), in the world who are working for the sewer rat agencies, are descended from the Aryans in the vicinity of Kazakhstan, as a result of the mass, forced conversion of a trading nation to Judaism, under the Khazar king, in the eighth century. It's the luck of the draw that he chose Judaism, in fact, because he was courting representatives of Islam and Byzantine Christianity, too, at the time, but the Jew was simply the cleverest debater.

That pagan king knew that in order to compete in the world market, he had to abandon paganism and adopt a mainstream religion. I guess it didn't work out for him, after all, because who's ever heard of the Khazars? This was documented by a Muslim scholar and historian, who accompanied the Arab debater to the court of that king. The Khazars' descendents in England were picked by London to recreate Israel, on behalf of the banking families, to colonize Palestine for them. Most of the powerbrokers in the occult/corporate order's Western tentacle, are Episcopalians and Unitarians, so the 'return of the Jews to Palestine' is bogus. Semitic Jews—the 'real ones'—are second-class citizens in Israel, of course, even the ones who moved there from other countries in recent decades.

I've known some Palestinians and many of them are also from the vicinity of Kazakhstan, strange to tell—lots of fair-haired, blue-eyed Armenians who are Christians and Muslims, for instance. 'Anti-Semitism' is a primary, ancient mind control scam designed to distract people from the real enemies of humanity: the patriarchal, occult/corporate world order.

Underlying realities like this one seem weird, because we're programmed to think they are. It doesn't seem strange at all to people who are no longer brainwashed, or otherwise compelled to police our own thoughts and ignore our intuition and gut feelings.

We've all been thoroughly programmed to ridicule or ignore whatever we weren't explicitly told about in school, church, cults or on The What To Think Network,' but the etheric realm is the one which is coming more alive, for many of us around the world these days, in the fast-expanding, orgonite network; a direct result/reward of our persistent efforts to systematically distribute orgonite in the environment and in our communities. Certainty is a rare commodity these days—have you noticed? It has to be earned through active discernment; it can't be absorbed.

The etheric realm is where we'll all be, when we abandon the physical garment, so why not familiarize ourselves with it a bit, right now? It's funny that the bad guys also know this and they believe that they'll remain here after death to continue to steal energy from people, (what else would parasites assume?), and their infantile, occultist secret societies reflect that assumption. I rather choose to develop certitude and faith, as a means of bursting from this material cage on the point of physical death and soaring into the next, more expanded and interesting realm, as one of The Operators' etheric helpers in this world. To parasites, this more restrictive world is the best possible one, but there's no accounting for personal taste. We can't understand a predator or parasite, without actually being one, so why try?

Empowerment and increasing accountability comes from that more enduring realm, not from the television teat, from flashy internet sites, from institutions, organizations or even from books. The accumulated, posted reports and consistent observations, on the internet, in several languages by self-motivated individuals over the past few years, gives this grassroots network credibility and induces more and more people to try their hand at environmental, atmospheric and social healing with orgonite. That's a sure-fire path to self-empowerment. Sure, it's not the only one but I challenge anyone to show me a better way, right now.

There are many, by now, who have broken away from the What To Think Network's relentless grip on our forelocks and have become genuinely self-empowered, accountable people. What many of our comrades tell us is that they've found their destiny and purpose in life, finally, through performing this selfless activity. When you produce grand-scale healing effects in the atmosphere and in your community, from your own gifting efforts, you'll directly grasp exactly what we're all reporting about on the legitimate forums.

What we know, by now, is that making and tossing orgonite is a trigger for the process of empowerment and that it's time for humanity to seize our collective destiny from the parasitic hierarchies, who have suppressed, camouflaged and pirated it, for six thousand years or so. If it weren't orgonite, something else would enable this new dynamic, because it's simply time for us to create a genuinely progressive civilization and thus achieve unity, prosperity and peace. The internet has been the precursor and enabler of that larger process.

The old prophecies of global mayhem and calamities, were always offered by genuine Prophets, along with the caveat that this would happen, if humanity didn't choose the high road, before the appointed time. It may be that we few thousand gifters and energy tossers are evidence that humanity did choose the high road, which is why there have been so few mass murder events, in the past decade or so, in spite of the What To Think Network's increasingly frantic, gloomy forecasts. Carol and I came to Florida, this time, to help the network disable the weather warfare apparatus in the region, in advance of this year's 'HAARPicane' season, which was promised by The What To Think Network last year, to be calamitous. Since all the big storms start in this region, we expect to get big results, from our surgical hits on land and sea.

When one is spontaneously serenaded by several Sasquatch, it's startling in a pleasant, uplifting way, to say the least. One's second thought is likely to be, 'Boy, if only the Pajama People could hear this!' The first thought, after the emotional slam and goose bumps, is: 'This is incredibly loud, spectacular and hauntingly beautiful!' Each time it happened for Carol, her daughter Jenny, and I that summer, we were outdoors and could look toward the little grove in the meadow, (the source of the sound), in the dusk or early dawn, but we never saw them, during the performances.

There's a genuine recording of a single, shorter Sasquatch call on http://www.coasttocoastam.com/audio/sasquatch.mp3, that you can listen to online, if you can stand to visit a popular disinformation site.
The first of the Sasquatch calls in that summer's serenades,

resembled coyote calls, only much louder, and when that commenced and more voices were added, various wild animals—usually deer, coyote and hawks—furtively crossed the ground, very close to where we were standing, between us and the pasture. These animals were looking toward the grove and apparently didn't notice us, though sometimes they came within a hundred feet or so, and it wasn't dark, yet. It's not unusual to get that close to deer or hawks, but one might spend a lifetime without being that close to a coyote. If it were happening to us in Africa, I expect we'd encounter some really big wild beasts, because there are plenty of those throughout the continent.

The meadow/pasture and grove is in the background of the first photo of the orgonite cloudbuster, in the 'Cloudbuster Construction' tutorial on http://www.ethericwarriors.com. The cloudbuster in the photo is on the spot where we often stood, to listen to the dusk and pre-dawn concerts, that summer. There was a disappearing chemtrail in the background, but the camera didn't show it.

The chorus seemed to be as loud as an air raid siren, and reminiscent of the recorded sounds of humpback whales. It 'felt' harmonious, with a lot of improvisation by individuals. We think there were five of them. We once went looking for them in that grove and we found some spots in grass there, which looked like deer had been bedding down, but of course the Sasquatch don't likely sleep within our solid reality. Carol indicated that she experienced something similar, when she was on a boat in Hawaii and several humpback whales swam directly underneath and sang; the sound literally shook the boat and was quite loud. She had tossed some orgonite, on the sly, right before and after that. The boat operators were environmentalists and would have been infuriated to see her tossing these healing devices in the sea. That wasn't Joan Ocean's tour, by the way.

The grove is a natural energy vortex, so we put some orgonite there to ensure its vibrancy and proper spin. It may be that the Sasquatch directly inspired us to gift that grove, of course. We don't pay much attention to claims that are made about Sasquatch and their origins, because most of that seems to be from runaway imaginations, silly channelers and cynical disinformants; titillation to get the reader/listener to buy into 'irrationalism.' Carol has the impression that the Sasquatch specie has been here for as long as we have. On our first gifting sortie in Death Valley, in June 2002, she had a spontaneous, clear vision of Sasquatch roaming that area, when it was verdant pastureland, before it sank down to its present level and later turned to desert.

Death Valley, four and a half years ago when we first visited, was quite dead; now it's a verdant garden, thanks to orgonite, including two cloudbusters, which we very illegally, (but lawfully), hid from the US Park Service there, far from the roads. We nearly got arrested on our most recent sortie there, when an off-duty park ranger, driving by, saw me hammering an earthpipe into the hard ground next to a sage bush, then reported us.

One of our intentions is to 'take back' metaphysical pursuits from the old pirates, who used a few hidden general truths and previously secret information, clothed in bizarre and irrational dogmas, as a highly structured, rigid and lifeless catchall to collect most of the intellectually-inclined in the West, who had become disenchanted with churchianity and were looking for more convincing answers. People who think, became disillusioned with the artificial dogmas and ideological baggage and began dropping out of the mainline churches during a period of relative enlightenment, a century or so ago. These alluring but confusing formulae had been developed and used successfully by the British East India Company, to undermine the Indian subcontinent in a previous, (eighteenth), century, in cahoots with crooked, greedy Brahmin rajahs.

'Cahoots' originated in the 18th century in America, origin unknown. I think it's a terrific word.

The Company set up a 'theological seminary' in Calcutta, then and the rajahs induced the populace to accept the graduates as priests. Hinduism and Buddhism weren't homogenous belief systems before that, any more than Christianity is, now.

That was probably the model that Alan Dulles later used to subvert South Korea with graduates of the New York Theological Seminary, (a CIA asset), after WWII. Sun Myung Moon came out of that mold, for instance.

The same, tired, irrationalist paradigm that generated Theosophy and lockstep Christian fundamentalism, was used to create the hippie movement later on, of course, which vamped a later interval of increasing awareness for our species and spawned the 'New Age' and 'environmental' movements. The present uplifting phase will not be subverted that way, because humanity, in general, is suddenly growing out of adolescence by now, and we're at maturity's threshold. This young but sober, grassroots and vital network's existence is evidence of that development, as is the internet.

I've been trying to help you get past the non-conspiracy theories you were marinated in, so please bear with me. This is all documented by reputable research journalists, and I recommend that you buy some books and get the information firsthand.

The reactionaries in North America—the demographic on the other end of the intellectual spectrum—who also became discontented with mainline Christianity in those days, were mostly absorbed by the plethora of mid-nineteenth century, masonry-derived American cults, such as Mormonism, Christian Science, Seventh Day Adventists, Pentecostals, Jehovah's Witnesses, etcetera, which all sprang up like meadow mushrooms, during that fertile time. The cynical Masonic origin of these quasi-Christian, 'anti-rational' cults, parallels the pseudo-Hindu/Buddhist expression of the irrationalist paradigm.

I repeat that the core spiritual teachings of Hinduism and Buddhism are empowering and pristine, as are the core teachings of Christianity, Zoroastrianism, Judaism, Islam and the scores of older, sometimes indigenous, revealed religions around the world. Clergy are responsible for the ideological corruptions that were later added to all of these, of course. I think we've grown well past the need to have clergy by now, but that we'll never outgrow our need for the Creative Word. I think it's time for intellectuals to stop throwing out religion, simply because of what clergy have done to them. If your priest buggered you when you were an altar boy, for instance, it's not God's fault; it's the priest's.

For that matter, if Sai Baba buggered you when you were a boy, (he was exposed for that six or seven years ago when scores of grownup—former victims—came forward with their public testimonies), it's because your folks, probably potheads at the time, lacked discernment and dragged you along with them to India; he certainly never represented the lofty precepts of the Bhagavad Gita, any more than predatory Catholic priests are qualified to intercede between you and God or Jesus.

America has a lot of distinctions in its short history, not all of which are flattering. The genocide committed by devout Christians against the Indians exponentially exceeds what the SS did to Jews and Gypsies, for instance. The racist version of slavery that was practiced in America is another case in point. Slavery elsewhere in the world didn't carry an ideological stigma, and slaves could purchase their freedom, though it's obvious that all people are now aware that slavery is unacceptable. I think that the origin of cultural schizophrenia in America started when Columbus claimed an already-occupied landmass for someone else.

'Christian fundamentalism' is the term that was invented for the modern, lockstep American phenomenon, represented by televangelists and an Islamic form of it was generated in London, a couple of generations later, and spread aggressively through the Middle East in the early 1900s, by the Pan Islamic Movement, in order to undermine the Persian and Turkish empires. Jalalu'd-Din Afghani was London's charismatic front man, then. After WWII, top SS men 'converted to Islam' and took the reins of the secret police organizations in all of the new Mideastern countries at London's behest, as has been documented in "THE HITLER BOOK", available from Franklin Press. The Zionists had put their British-trained Zionist counterparts in Israel, by then, but there's no clear distinction between the Mossad and the SS. Their common intention is to start World War Three, but we believe they'll continue to fail.

Espionage is a pretty small community and it was during that postwar period that 'A MAN CALLED INTREPID' was published and indicated that there's really only one espionage agency in the world. London conducted the Cold War, using the US and Russian military machines, as a way of consolidating the communist and fascist regimes' control of all developed nations. I think the really big surprise, that all of this has been orchestrated from a remote area in the Gobi Desert all along, is about to be sprung on the world. Let's see whether I'm correct but, meanwhile, just put that notion in your 'wait and see' file.

We're living in a day when all hidden things are being uncovered. How much of the previously-hidden horrors in this world are you prepared to contemplate and then neutralize now?

When the psychics in the network really get to digging into the available etheric intel, they find connections between all of these overtly conflicting hierarchies, which mostly lead to the Vril Society, the top of the Illuminati pyramid, (read: dung heap), and since August 2005, to the Beijing and ancient Gobi hierarchy. Most of this hasn't yet been documented, because there is nothing in the public record about it, aside from the fact that Clinton, at Bush Senior's behest, gave China our patent office and the Panama Canal; also built the Chinese Army a container port in Los Angeles.

The German and Jewish research journalists, who are published by Franklin Press, have provided the most documentation about the present activities of the SS, but they don't dig any deeper than that. The Nazi regime was the brief surfacing, in human society, of this deeper Vril hierarchy, which the SS represents. That organization is mostly hidden in corporations, Western spy agencies and in the secret police organizations of Israel and the Islamic countries. Did you believe that it's coincidental that the Zionists are treating the Arabs in Israel exactly the same way that the SS treated the European Jews? Many Jewish Israelis are acting on their consciences, by openly criticizing the bloodthirsty Zionists now, and that takes a whole lot of courage, as you might imagine.

Reading the reputable Jewish and German research journalists, whom Franklin Press publishes can help you break down the popular misconceptions about recent history, which the What To Think Network have planted in our heads, I think. It's curious that more of the reputable conspiracy researchers haven't delved into the deeper, occult material. All of them pretty much stopped hunting, when they reached the ends of the money trail in London, the Vatican and Switzerland. Beyond that, it's not about money any more, of course; it's about stealing energy, instead of just accumulating goods and governments. Their Luciferic ideologies are infantile, but they have to be reckoned with, even so, because their adopted life purpose is to destroy the rest of us, and they control most of the world's physical resources.

Discernment is kind of like pregnancy. We're either discerning in the moment or we're not. It's best to remain discerning, of course, which is to say we need to keep our minds open, but free from prejudice, belief and denial, when we're seeking and acquiring facts and confirmations. None of us were taught detachment, any more than we were taught to be parents, so we all have to learn on our own. In the emerging paradigm, schoolchildren throughout the world will receive training in discernment and parenthood, of course. Until World War Two, one could receive good, classical training in discernment in many universities, before corporate and National-Socialist funding favored Marxism and other infantile educational paradigms.

I'm hoping you'll do what most of us are doing and attempt to discern every new thing that crosses your path. The truth was never popular, but truth does, indeed, set us free. Achieving certainty and faith requires commitment, a rare commodity. If you want to tell about reality you'll invite scorn from the PJ folks and the What To Think Network's editors and censors. Anyone can be superficially committed to fundamentalist Christianity or Theosophical cults now, though, because there's no genuine risk involved, unless you practice them in China. How many were willing to commit to the nascent Faith of Jesus at the height of Rome's empire, when there was real risk? Not many, actually, and Theosophy-based cults always had the enthusiastic backing of the patriarchal secret societies, whose top, hidden members have been operating under the receding paradigm for centuries and even millennia, so there was never any risk in embracing or proselytizing irrationalism. Christian fundamentalism and Theosophy are still promoted as 'revolutionary', to chumps who are led to believe that 'the other one' is evil incarnate.

> The origin of 'chump' in Britain is 'the thick end of something, especially loin or lamb or mutton.'

That certainly lines up with my impression of the average born-again Christian or new-ager and whom do you reckon is feasting on these goofy muttonheads? There's an awful lot of both among the PJ folks these days, also quite a few among the predators and that fact shouldn't be ignored.

The world order has mainly waged psychological warfare on humanity, throughout recorded history. The conquest of India is a case in point. The British Army was pitifully small, when the British East India Company took over the subcontinent. Those gouty, corporate parasites, in London took it over with subterfuge and their own 'Hindu' clergy. The army mostly came in after that to enforce colonial rule, on a dispirited and manipulable population.

The rajahs themselves were induced by bribes and extortion, to cut off the right hands of all the weavers, (India was renowned for fine fabrics but the Brits wanted that market), and to make subsistence farmers grow poppies instead of food, before and after the British Army arrived. Opium was essential to London's material exploitation of China, but that agenda was cut short by mass starvation in India, for which a somewhat conscientious German, Queen Victoria, disbanded the British East India Company in disgust, early in her reign. I think the company had 'liberated' enough of South China by then to buy opium there, for which they traded gold in Hong

Kong. Meanwhile, historians ignored the deeper currents in China, until the present. Who would have been able to predict MalWart and China's control of the US Economy and the Panama Canal, just a few decades ago, for instance? The PJ folks' ire, before the surprise debacle, was directed at Japan and the Arabs.

In order to grasp even the simplest truth, we have to empower ourselves first—don't you agree that this bears repeating? The magical inner dynamic that occurs, when we perform the selfless service of healing the environment and society by distributing cheap, homemade orgonite, genuinely empowers anyone. Reading substantive books consolidates this achievement, because knowledge and certainty depends on acquiring facts and usable information. I don't know of an easier or more painless way to achieve that, in fact, and I'd been looking for empowerment most of my life, previously. We never actually get enough self-empowerment, just like we never get enough healing in this life. The other side plays such a deep, deep game of deception, that we get endless opportunities to sharpen our discernment skills and to heal the effects of their exploitation.

An old trick of the occult/corporate hierarchies is to infiltrate, then seize control, of every progressive movement. They've failed in this case, so far, because we're not centralized and the old regime can't comprehend the nature of grassroots movements, or even the nature of spontaneity or human intuition. This is because intuition is a heart function and these occult/corporate rebels have shut down their hearts pretty thoroughly, by consistently choosing to exploit, rather than serve others, in the moment.

If we had agreed to establish an arbitrary hierarchy, in this network, (many offers were made), they'd own us outright by now, and you probably wouldn't have heard of orgonite or wouldn't have taken it seriously, if you had heard of it. The other side would have overlaid our reported accomplishments, with a boatload of saccharine Theosophical regurgitations. That would have made this network just about as fascinating and viable as maggot-ridden roadkill.

Genuine personal empowerment is impossible without deeds, and if one is trapped in the mind, which is what both the What To Think Network and the irrational cults are designed to ensure, one can't access one's own discernment, which is mainly a heart function. One is conned into believing that squirts of endorphins in the programmed brain, triggered by key words and phrases, will be acceptable substitutes for the upwelling force of the heart's recognition and acceptance of a larger view of reality. The other side is pretty resourceful with those endorphin squirts!

In the process of accepting others' programming, one has also let these parasitic programmers into one's heart. The unwitting victim's cordial invitation to parasites is the other side's clever sucker punch; the hook hidden in their bait. Nobody can force his way into a human heart, of course. One has to be invited and the Friend will simply leave, the instant that the enemy is invited inside.

The new-age movement and religious fundamentalism both have this programmed-titillation feature and both are creatures of the old order. Lyndon LaRouche's comment about Christian fundamentalists is, 'They've rejected eternity for a bad infinity.' LaRouche owns Franklin Press, by the way. I don't endorse his politics, or his organization, but I admire his intellect, personal accomplishments and self-sacrifice. Those authors document extremely well, so their factual findings are useful. Most of them are independent; not LaRouche's sycophants.

The John Birch Society also has a very fine reading list and sells books. The Rockefellers funded that organization in the beginning, in order to further polarize America, but Gary Allen was one of their authors, twenty years later, and there are other reputable authors whom they publish. Gary Allen met his end shortly after the Birchers published his expose about the Rockefellers. The 'Christian-new-ager' fake conflict is paralleled by the original Birchers', 'communist-capitalist' puppet show. I think it's a modern miracle that the Birchers, at least, eventually figured out that communism and National-Socialist capitalism are joined at the hip.

Eustace Mullins has published several excellent and diverse conspiracy books, as has Jim Marrs, who identified three of JFK's assassins, and got their written confessions, which the What To Think Network opted not to mention, nor did anyone in the press show up at the press conference Marrs held in the early 1990s, in which those shooters wanted to publicly confess.

Our friend, Dr Len Horowitz, and a few other research journalists have published watershed books about modern biological weaponry that has been deployed against entire populations, in recent decades. These strategies account for the plethora of new chronic illnesses and also for millions of untimely deaths. They identify Dr. Gallo as the inventor, not the 'discoverer' of AIDS, for instance. If someone like Gallo and Kissinger get Nobel Prizes, what does that say about this dubious distinction? Horowitz exposed the Bird Flu scam, essentially ending, it last fall. That scam sold an awful lot of our zappers, but I wasn't sorry to see it end, of course, and Doc H started buying Carol's orgonite healing devices for distribution, before that.

I'm skimming the surface of available, reputable conspiracy information sources, of course. If you want to consolidate your discernment successes, you're going to need to read some of these books. If you had read some of them before you found orgonite, you're a couple steps ahead on the learning curve, because your mind's eye is relatively free from the blinders of the What To Think Network.

This global gifting network is quite diverse, in terms of religious, cultural, national and racial backgrounds and affiliations. When we get together, here and abroad, there's genuine love, humility, self sacrifice, empowerment and acceptance among us. Our bond is the wonderful, invigorating, healing work we've discovered and are accomplishing singly and together. It's a spontaneous, synergistic heart bond among us, not an ideological one.

This level of social harmony may be rare these days, at least in the West, but it's also a precursor to the general acceptance of this way of living and I think it will contribute significantly to genuine, dynamic unity eventually. If our bond weren't based on deeds and shared accomplishments, it wouldn't be this deep.

'Common sense' and 'conventional wisdom' have always been oxymorons, in the schizophrenic West, but we hope to change that, by simply distributing more and more orgonite.

When traditional metaphysics mavens, almost all of whom are Christian/Jewish-derived, Theosophy-affiliated Europoids, get together, there's a lot of convincing talk about selfless love, humility, empowerment and acceptance, but what's really going on among them, is that they're each either sycophants, genuine seekers, (passing through), or are exploiters, looking for an advantage over the rest. All of them seem to hate revealed religions and, by extension, our Maker, strange to tell. Because they represent the catchall—the ones who broke out of their hypnotism by the What To Think Network or churchianity, but lack social conscience or a desire for genuine spiritual progress—they attempt to make up in decibels, conformity and numbers, what they lack in genuine integrity. This can be quite distracting, for the unwary and in addition to that veil or barrier, the new-agers tend to proselytize quite aggressively, too, as any cultists do. Preventing them from seizing control of the nascent orgonite network, was my full-time job, for the first three years or so, and I mean several hours per day spent with personal correspondence and posting.

At times, they were like a human wave, and quite well orchestrated from behind. Aggressive pot addicts made up the vanguard, not unlike the way the Chinese threw under-equipped, pot-inebriated soldiers in human waves, at the Americans, in the Korean War. Creepy; I'm glad you didn't have to witness it. I never insulted anyone personally, in all that time, because I was being set up to do just that—by being reactionary I would have lent some credence to their cynical, saccharine, 'Love and Light,' aggression. I got it done by clearly stating the issues and the dynamics involved and, after awhile, balanced and sober people came in, to exemplify the network.

Many moved through that artificial new-agey-stagey paradigm on their way here, and our challenge is probably to keep the good stuff they managed, (against the odds), to glean from their experiences with those people, and just leave all of the sticky, smelly, smiley and saccharine, mind-numbing programming behind. The useful material that one might have gleaned, during that dark passage, had been inserted into the programming or allowed by cynical handlers and programmers, in order to give their artificial paradigm credibility, of course. D Bradley was one of the premier new-age programmers, until he finally recognized who was pulling the strings and started working, from the outside, to reverse that agenda.

Everyone has an innate sense for what's real, but most people are conditioned to believe that it's okay, if their spiritual leaders have reprehensible characters and exploit others. It's really not okay to be a hypocrite and we shouldn't subject ourselves to clergy types, ever!

What cultists believe is mostly in their heads—as I mentioned—not in their hearts, and their professed beliefs are polluted with institutional agendas, which induce arrogance and isolation. When a sincere person unwittingly gets drawn into a cult gathering, he or she usually wanders back out again, fairly discouraged and a little more cynical than before, unless there's a fire of genuine, passionate search smoldering in his or her heart.

When the more vital seeker eventually sees what we're all doing, in this network, it's a breath of fresh air and he or she can't wait to get a gallon of resin, some metal waste and a few crystals and directly experience what we're all talking about. The hope of genuine empowerment and direct knowledge can be pretty compelling, after all.

Gifting causes latent psi talent, intuition and other etheric gifts to develop rapidly and practically. That's something that isn't available from human hierarchies. It even generates opportunities to become more materially independent. Earning a livelihood is an essentially spiritual exercise, after all, and one's spiritual powers, hence one's potential value in the market, are enhanced by selfless service. This is another expression of the fact that real freedom depends on individual, economic sovereignty.

The problem with reporting subjective experiences like the Sasquatch encounters, is that it so severely challenges some personal paradigms, that I'm compelled to reassure readers, very often, that it's not necessary to believe or disbelieve what I'm writing. Just read it without prejudice—reserve judgment and move on! I'm laying out some pretty lengthy observations about the process of discernment, because the more discernment we can develop, the more of the finer realities, like Sasquatch, aliens, dolphins, underlying energy dynamics and perhaps even the human soul, will become apparent.

A few days before we arrived at the Washington location, we spent the summer solstice night under gigantic fir trees near Panther Meadow, high up on Mt Shasta. We were serenaded, from sunset to sunrise, by a Lemurian choir, as Carol and I watched their flashy ships darting around the starry skies and Carol conversed with them directly, though telepathically. Neither of us could sleep at all, because we bedded down where the black ants crawled on us all night. That might have happened by design, but we certainly weren't bored and we preferred to stay awake and savor the experience.

We later saw a bright blue Lemurian craft slowly descend to that property in Washington, as we were approaching home late one night, from two or three miles away, so it seems obvious to us that the Lemurians, who are humans that also live among us hyper-dimensionally, are affiliated with the Sasquatch.

Dorothy West, another of my teachers, who is a Seneca Indian elder and a member of the very old Doran order of Druids in Wales, spent a day in Panther Meadow, on Mt Shasta, in the early 1970s, on her way north from Los Angeles. She was greeted by people she presumed were nice, oddly-dressed hippies, who invited her to picnic with them. After the picnic she was invited to take a little nap with them in the meadow, then they all woke up, and she was escorted through a very tall tunnel into an identical meadow, where she spent some time chatting with them. Later in the afternoon, she was escorted to the parking lot and left. Much later, she discovered that these were Lemurians, and that they'd taken her into their world, which is actually hidden in our world.

Luis Ortiz, a gifter from Mexico, with whom Carol and I did some experimental work in Oregon a few years ago, had been trained in some of the old Yaqui metaphysical arts, during his youth. When I told him these two accounts he said that when he and his brother were riding a bus to Portland, Oregon, several years before, the bus broke down on the highway near Mt. Shasta and the passengers got out to walk around. Two tall, bearded white men in robes approached Luis and began talking to him. Shortly, Luis' brother came up behind him, and he turned around. When he turned back to the men in robes they had vanished.

D Bradley told me that he, his wife and their lawyer explored Pluto's Cave one day, north of Mt Shasta. That was after DB had turned against his former masters and was in the process of exposing their genocide agenda, on lecture tours with Ted Gunderson and David John Oates.

They went pretty far, along that ancient lava tube, which the Indians allegedly once used as a convenient pathway into the Cascade Mountains, farther north in Oregon. The three hikers discovered what appeared to be Atlantean ruins in the cave, which is quite huge in places, but fairly level, and never far below the surface. In the car on the way back to town, Count Ste Germaine materialized in the backseat and told him, 'If you ever come back here again, you'll be killed,' then disappeared. The other two in the car also saw and heard him.

Carol and I found and gifted the ritual site in that cave, where DB told us that the Church Universal and Triumphant's upper hierarchy, in the presence of that materialized parasitic alien, (disguised as a handsome French noble from the1700s), conducted human sacrifice rituals and buried the bodies. The ground in that spot is spongy and has the specific, sweetish odor of decaying flesh. Elizabeth Claire Prophet founded that Theosophy-derived cult, and channeled insipid metaphysical books, that the alien disguised as Ct Ste Germaine allegedly dictated. DB and Carol tell me that this is a CIA front organization, which has its own regional espionage and assassination network, in the Pacific Northwest. It's easy enough to see that, when one gifts anywhere around Mt Shasta or Yellowstone National Park. We always took our pistols along, when we gifted in those regions, because we often encountered people who apparently were looking for opportunities to harm us.

He sent us a long list of other, hidden ritual sites in the Mt Shasta region, which we also gifted that week, sometimes to our peril, due to the unusually aggressive surveillance, by those cult members. We made sure that the creepy-looking people, whom we encountered in those remote places, saw the pistols on our belts.

All of this is in the journal reports, in more detail, but I'm just mentioning this little sampling, to paint a picture for you of what we, and many others, have been doing here and abroad, during the past five years.

I won't often name alien races or species we've interacted with, because all of those names were also turned into mind control triggers. If you haven't witnessed a group of clueless new-agers eagerly discussing Wingmakers, Andromedans, reptilians, Atlanteans, Pleiadians, Draconians, Annunaki, et al, you can do well without that bizarre experience. That hasn't been done successfully, in the case of the Sasquatch, Lemurians or cetaceans, thankfully, but it probably will be soon, I'm sorry to say.

Hopefully, telling about our personal experiences with non-humans, will induce you to be a little less judgmental, credulous and skeptical rather than more. I'm certainly challenging you now to open your mind up. You already bought the book, and I'm not trying to sell you anything else. It's fun material, obviously, so what do you have to lose by having an open mind? We're not an organization, so nobody's going to accost you or pick your pocket.

As James Hughes says, if we can stay away from judgment, belief and denial, while we're examining reality, our search will be productive.

> Prejudice: an opinion made without adequate basis.
> Judgment: an opinion or conclusion.

My own faith teaches that love, under the circumstances, can be as grievous a barrier to discernment as prejudice, because it cancels detachment just as effectively. In order to examine new information effectively we need to be detached.

At this point, you've certainly got no basis to deny my subjective reports and the obvious transformations that have taken place in the environment and in society, due to thousands of people's orgonite distribution efforts, which have given credibility to our discovery, at least, and made us reputable. If you haven't conducted an independent search, you've also got no basis to deny my conclusions about conspiracies, because the What To Think Network and your university professors won't debate the subject of conspiracy; they simply scorn the subject.

There's a good precedent to our discussions about aliens:

Dr Reich invented his cloudbuster as a weapon to drive away alien ships that were being used to slowly kill everyone in his

lab, Orgonon, near Rangeley Maine, in the late 1940s.
In that case, the lab was a very big orgone accumulator and the alien ships were directing concentrated, deadly orgone radiation into the lab, which condensed the energy even more and then radiated it outward. Deadly orgone radiation, (DOR), is the opposite of life force. It's the energy matrix of nuclear weapons and the other deadly technologies of this old world order, which is why all of that is easy to disable with orgonite, which transmutes deadly energy back into life force, perpetually.

His cloudbusters later were used to end droughts, in North America, and to even temporarily reverse the desert in Southern Arizona, in 1953. That was during a time when atmospheric nuclear blasts were being used in seventeen US States, (I bet you thought it only happened in New Mexico and Nevada then), to generate catastrophic weather on the continent, and public outcry eventually ended atmospheric detonations. Note how our network showed up during a time when the 'next level' of weather warfare had been established. Can you see The Operators' hands in that happy development? This is a pattern that's been consistent throughout human history, please consider. If that weren't so, our planet would probably have been turned into a cinder or a big, festering boil by occult/corporate parasites, before now.

Very few people are willing to openly discuss Dr Reich's defeat of those predatory aliens, but lots of people are able to freely discuss his weather modification successes, at least now. Back then, I don't think the PJ folks were capable of noticing even something obvious like that. Reich was always considered to be a man of personal integrity, by the people who knew him, including Sigmund Freud and Albert Einstein. Both Freud and Einstein later turned vehemently against him, of course, because his findings potentially turned their own professional paradigms upside down. Reich had a curious way of inducing a sharp distinction between the sheep and goats, among the people who knew him.

Since some of what I'm telling you can't be documented, (it's subjective, after all), there's also no basis on which to deny it, as I've said, but others' corroborating accounts can be assumed to be empirical evidence, as long as they're also reputable people, as Carol and I are. I've invited as many of these folks as I know to post on www.ethericwarriors.com.

Intellectual honesty has been useful, when the psychics in this network pool their resources during the intel-gathering sessions. As a rule, when one skilled psychic sees something substantive, it will also be seen by others from their own perspectives, which often lends a pretty comprehensive view for everyone. The exception is when the psychics' predators and ritualists on the other side throw false imagery at one or more of our psychics, but this ultimately just induces our folks to develop their skills more and to look more deeply into the targets. They invariably find consensus that way. The false imagery is usually seen as two dimensional and static, just like a photograph. Looking harder into that imagery usually enables the psychics to see what the other side are trying to hide, which has depth and often movement. The psychics help each other, too, with timely suggestions and observations.

The enormous, international Monarch Program, (widespread, systematic, trauma-based mind control among millions of Whites and Asians), was created after World War Two, based largely on what was done at Tavistock Institute in Great Britain, using German prisoners of war as guinea pigs. Sigmund Freud was in charge of that. His daughter runs Tavistock now. Coming to terms with the reality of such a vast mind control program, helps us understand why millions of new-agers and fundamentalists are repeating looped Theosophical and neo-

Biblical platitudes, but can't get their personal lives in order. One of the oldest Big Lies, is that human, patriarchal institutions have the authority to make spiritual judgments and even create ideologies. How did we buy into that confidence game? Do you think that the Dagon symbol on fundamentalist Christians' tailgates is amusing? I do. The worship of Dagon, the Babylonian fish god, in ancient Rome at the time of Christianity's appearance, required human sacrifices. The Pope and his bishops, like their pre-Christian counterparts in the Temple of Dagon, wear the mitre, which is a fish head with an open mouth, on their own heads, during their primary rituals—a vestige of the old costume worn by the Pontifix Maximus of the Dagon cult, in pagan Rome.

Mel Gibson's slasher movie about the torture-death of Jesus Christ, showed how Mithraists treated their enemies and sacrificial victims. Mithraism, which was the Roman soldiers' religion which carried the image of the 'Christian' cross into battle, long before Jesus lived, later became one of the main ideological underpinnings of the all of the Christian churches, in the West.

I personally agree that Jesus survived that event and moved away with His uncle's help, but that's also just my belief, based on what I've read about Rennes le Chateaux and Rennes les Bains, though it's also a belief that's also promoted in the Qur'an. Apparently, Jesus was married to Mary Magdalene and she and their children were taken to Southern France, after the crucifixion. There's an awful lot of circumstantial evidence, but the best evidence is apparently still hidden, perhaps within the Vatican. To a rational person, this doesn't degrade the Divine station of Jesus Christ any more than Moses, Muhammad or Abraham were compromised by having families of their own. The dualists, who influenced early Christian church doctrine, are responsible for the common patriarchal notion that sex is somehow evil. The

Bible's misogynistic, unauthorized editors are responsible for women being perceived as evil.

Does a human institution, which uses regurgitated satanic imagery and calls all of that 'sacred', deserve to tell you what's real and what isn't? Do the clergy of that institution really serve as intermediaries between you and God? That sort of unlawful authority corrupts people, obviously.

Abraham Lincoln liked to confound his very-devout Christian critics, (in the days of the 'righteous' Indian massacres and racist slavery), by saying, 'If I call a tail a leg, how many legs does a donkey have?' He knew that shallow, irrational people are conditioned to say, 'Five,' and he no doubt then enjoyed stating the obvious, 'No; four. Calling a tail a leg doesn't make it so.' It's fun to confound cultists—try it!

According to a Persian saying, 'An hour's reflection is better than seventy years of pious worship.' Critical thinking is the only thing that can help one break free from the shackles of the What To Think Network and the older ideological traps. One shouldn't ever accept what is repugnant to the rational mind, I repeat. The notion, in the face of massive evidence to the contrary, that human history is chaotic rather than orchestrated, is repugnant to the rational mind, for instance.

A functional level of skepticism is healthy, of course, and the only 'judgment' that counts for the private individual in this life, is personal discernment. Judging others has never been a personal prerogative, nor will it be, in my opinion. Judgment of other individuals is a function of our Maker, and in cases of overt criminal behavior, of lawfully-operated courts; always has been and will be. That's just not our job, unless we're elected judges or jury members, at the moment, and we're deciding the physical fate of an alleged criminal.

Most folks, even people who strongly disagree with my assumptions, easily agree that my offerings are harmless and entertaining, at worst. I'm obviously not selling anything, because my journal reports, in exhaustive form, are still available for no charge, on the internet.

I'm banking on the assumption that the larger view of reality that we're now glimpsing and attempting to describe to you, will become the common paradigm, before long, as all of our specie continue to relatively wake up. As more and more people see these strange, new phenomena, there won't be an excuse to ostracize the few of us who have been seeing them for years, and the What To Think Network will eventually expire for lack of subscribers, after all, so why remain vested in it?

We'll look back on the way that the What To Think Network have held sway over the PJ folks' sensibilities, the way we look back with amusement on the massive pomp and display of crowded Nazi rallies, in 1930s Berlin, or the mandatory parades past enormous portraits of Lenin and Marx.

One moves toward personal faith and certitude by first committing to the process of gaining faith, and this is a process that institutions simply can't get involved in; it's always a private, intimate process and it's a function of God's grace, not 'karma.' Grace is something we desire in our hearts, which implies humility. You won't meet many arrogant people among this network.

If you will discover, as Carol, I, and many others have lately, that reality is much more interesting than can be described by the best science fiction or fantasy writer, then we'll be on the same page. The What To Think Network and just about every other depressing, human institution these days, are fully committed to prevent you from empowering yourself this way, of course. They induce their thralls to embrace a bizarre, schizoid mix of programmed self-loathing and arrogance.

The more aggressive, fearful Pajama People will try their best to dissuade, and then ostracize you, if you feel a need for their validation, of course, because reality's larger implications make them intensely uncomfortable. Life has always been like this, and probably always will be, in relative terms. Today's PJ person knows more, for instance, than the intelligentsia of past ages did.

Did you know that this nifty word, 'intelligentsia,' has Russian roots, by the way?

If you were born and raised in a third world country consider yourself fortunate, because the occult/corporate world order never expended much effort and resources, besides sending penurious, but fanatical missionaries, in the previous two centuries toward controlling the thoughts and behavior of your people. That means you're potentially more capable now, of critical thought and discernment, than the average Westerner or industrialized Asian is. If you've spent a lot of years in the West or were educated by missionaries, you might feel that your origins don't count for much, but the tables are turning now, as our African co-workers are steadily discovering and demonstrating.

Back to the Sasquatch:

Carol and I took a long route back from the grove the day we gifted it, through the woods and around the property to the south. At one point we both heard some rustling nearby and briefly smelled the characteristic scent that only Sasquatch give off which, frankly, is pretty repugnant to humans. Carol said that this was their way of showing us that they were still around—a little encouragement from them and maybe a 'thank you.' Another time, years later, we were tracked by a cougar, in a remote part of the Bitterroot Mountains, during a tough hike on our way back from the site of some Celtic ruins, and that felt a lot different. That day, we were armed, because the remote vortex we had gifted was a favored murder ritual site that the Jesuits established in the early 1800s, shortly after their arrival in the region. There were Celtic artifacts there until 1949. When the Jesuits first set up shop, in a valley farther north, they made a beeline to the Celtic ruins, so they apparently knew about them already.

A copper miner, who came to own the property, bulldozed the row of inscribed dolmens into a trench in 1949, and buried them, in order to keep his mine from being shut down by archaeologists. The man died in recent years, then Carol visited there with another reputable psychic, who was hired by a surviving companion of that miner to find the stones. The miner buried them in a panic, because there was a lot of discussion being generated in the regional newspapers in those days, about the site. He needn't have bothered, because nothing that challenges the Columbus paradigm ever gets mentioned in archaeology publications, of course. What a tragedy! Lots of folks had visited the site and described the Oggam inscriptions on the line of dolmens, on that ridge, so their existence wasn't in question.

It took an awful lot of orgonite, on three successive visits, to turn that big, nasty vortex around, after the Jesuits had been pirating its strong energy with regular human sacrifices, for a century and a half. Carol's first visits were quite dangerous and disorienting, in fact. Dr Steve Smith, Dooney Weise, Ryan McGinty and I went with her on Carol's third sortie there, and we finally got enough orgonite into the vortex to get it spinning the right way. It was a bizarre day for us all, nonetheless. The whole region lit up after that, and lots of people started going to the area again, to experience the beauty of that wilderness.

We'd camped the previous night at the summit, above the snow line, where it was quite windy. I luckily brought along our half-scale, three-pipe cloudbuster, and the wind stopped immediately, when I set it up. It remained calm all night, but a grizzly bear came to visit and left a big, stinky calling card, right next to the Jeep, as Carol and I were sleeping in the back of it with the tailgate open. On her previous visit, with another witch friend, they came upon two bloody sleeping bags that apparently had human remains in them. On their way back to the campsite that day, the bags were gone. She suspects that was a prop, put there by the Jesuits or one of their affiliate sewer rat agencies, to stop them from gifting the vortex. She had her Glock Nine out all the way back, though, because a cougar tracked them very closely on a parallel course, growling often. What a gal, eh?

My wife has a personal history with some cougars. When she was living in St Maries, Idaho, which is where I met her, a cougar gave birth to a cub in the vacant, overgrown lot behind her property, which was right in the middle of a residential neighborhood. The cougar chose the mown spot, which Carol was using for her personal rituals, in fact, and the animal never attacked anyone, though some of the small dogs in the neighborhood disappeared, over the next eight months.

Toward the end of that period a professional hunter was called in, but he failed to find the cats, which returned to Carol's spot after he and his beaters had given up and left. 'Juju, Bwana!' The cougars left not long after that. Carol's pretty good at keeping secrets. That was a few years before I met her, and it was a few more years after we met that she told me about that. To her, this is as unremarkable as having interacted with a dog or cat.

We got a puppy in December, (got tired of the feds breaking into our home in the middle of the night), and Carol's telepathy provided me with no end of fun with this dog, a white German shepherd we named, 'Neha', (Nay-ah). Watch dogs are wonderful assets, when one is under constant surveillance by the unlawful, quasi-governmental sewer rat agencies like the NSA, CIA and FBI, because neither physical nor astral spooks like to be exposed by barking dogs, especially when they can be efficiently assaulted in the etheric realm, which is more real than the physical one, after all. That's apparently why they killed her a few weeks ago. When these human predators get near orgonite they're off balance, which makes it even more fun to smack them etherically. If you're lucky, you won't have to contend with these Homeland Security Abominations. Their resources are getting stretched pretty thin, lately, by this expanding network and by other genuine patriot associations, of course, so maybe you'll be lucky and they won't try to intimidate you away from your healing efforts.

Once, also before I met her, she was riding in a pickup, with her previous spouse, and as they passed an owl, which was lying on the side of the road, apparently stunned, Carol felt the owl specifically call out to her for help, so she told her hubby to stop. She walked back to the owl, which looked steadily into her eyes, and picked it up by holding the two wings with her hands, out beyond reach of beak and claws. The owl remained passive, though its head swiveled so that it could maintain eye contact with Carol. Right after she set him, upright, away from the road he flew away. If you've never seen an owl take off from a standing start you're missing something spectacular.

We often take walks in cemeteries, so Carol can commune with the owls, which are usually expecting her. Living with this woman has taught me some profound respect for the nature of animal intelligence. I'm finally starting to get why the Indians called them 'brothers.' 'Dominion over the earth' doesn't imply innate superiority, any more than the janitor in a school is less exalted than the rest of the staff. I feel sure that homo sapiens has dominion in this world, even though there are obviously some more advanced species who also consider it their home. I think clergy have added the arrogant value judgments. The Prophets, themselves, have always been the Essence of humility, after all.

I hope you, too, will develop an appreciation for the pure intelligence of all creatures and for their constant, intimate interaction with us. This is one of the bounties of our physical existence. Gifters soon come to realize that birds and animals are giving us signals and hints about where to place our orgonite, because they have a vested interest in our success, just as the elementals do. The new weaponry of the occult/corporate order extends its poisonous influence through many worlds, apparently, not just the 3D one.

If you're a cat lover, you already know what I'm talking about, of course, and night time city gifting sorties will hold some psychic treats for you, because the cats are out in force then, and they seem to want to help us get it done efficiently. When we lived in Moscow, Idaho, we were often visited by a lovely black and white cat, who sometimes came into the house without the use of a door or window. I don't think she was trying to impress us; she just wanted some affection, a snack and a warm, dry place to take a nap.
Another time, during the summer of 2001, in Washington State, Carol was mowing the yard and found a small clump of fur that smelled strongly of Sasquatch. She came to tell me about it, but when we both immediately went back to the spot, the fur was gone and so was the smell. That was a personal gesture from a Sasquatch to Carol, of course.

My only 3D view of a Sasquatch was from inside the camper on the back of the pickup, looking forward, through the pickup's windshield. I saw what was apparently a mature male, running from right to left, right in front of the truck's hood. I ran out the back of the camper to see him, but he'd gone, of course. Then I realized that the worktables in front of the truck would have made it impossible for you or I to physically move across that space. How solid do you reckon those tables actually are?

Carol's longest view of a Sasquatch that summer, may have been of the same male, who ran smoothly from right to left across the long meadow that lay between us and the grove of trees, which was a quarter mile or so east of the lawn. It was only when we walked through the meadow later on, to gift the grove, that we realized that it's impossible to move smoothly across that space, because the surface beneath the tall grass was made up of deep, overgrown and badly plowed old furrows, which the deep grass had hidden from view.

Another time, Carol stopped her car along the quarter-mile-long driveway to watch an adult Sasquatch, walking in the woods with his back to her, holding the hand of a very young one. When Carol stopped the car, they also stopped, turned their heads to look at her, then both Sasquatch faded from view in a couple of seconds. It's not hard to see why a Sasquatch has never been captured.

Chapter Three

Discernment Hound

Danion Brinkley, whom we had the pleasure to meet in Spokane in August 2001, feels that the internet and cell phones are simply stepping stones leading to mass telepathy, which the more advanced species in this world, along with many animals, enjoy, so we should use all of that affordable technology, not shun it. There's no real glory in being a Luddite, and shunning technology won't move us closer to understanding species like Sasquatch and dolphins. In fact, shunning technology, pretty much guarantees that we're not going to get our physical bodies close enough to have these encounters, nor will we have any discretionary time to spend examining stuff like this, if our first efforts are spent on mere survival.

If you haven't read any of his books, I hope you will, because he provides a fresh glimpse of the next world, that can help you get clear of any residual, mind-crushing churchianity and Theosophical dogmas, that might be chained to your ankle. People who choose to believe in devils, are as uneasy with his liberating disclosures, as people who choose to believe in reincarnation are, but he's so engaging, funny and genuinely loving that his presentations are pretty easy for anyone to assimilate. Too bad his next-world travelogues don't have a slide show.

I know that I anger everyone who cherishes the programmed, irrational belief that human souls move through a depressing, linear progression of physical bodies, but on the other hand, folks who proselytize that artificial paradigm are generally duplicitous and untrustworthy, anyway, so winning an ideological p!$$!ng contest against them would be a hollow victory. All I can say to them is, 'I'm sorry my karma ran over your dogma.' I'm appealing to the smaller, broader-minded demographic, who rather put a premium on intellectual honesty. Truth has never been popular, of course, but those of us who constantly seek it, find it more desirable, ultimately, than sex.

The most one can say about the soul is that it's essence is unfathomable and that it's infinitely multifaceted, I think, and that the aggregate of souls are individuals, sure, but in the same context as waves on the ocean or leaves and flowers on a branch. The incessant attempts of faux-spiritual ideologues to stuff all of that into a featureless, one or two-dimensional concept, just feel repugnant to me, and I wouldn't mention it at all, if the What To Think Network weren't vigorously promoting it as 'the alternative view' to churchianity, which is just a wee bit more schizoid than 'irrationalism' is. The old, worn-out reincarnation scam seems as silly to some, by now, as the squeaky, 'I am God!' rantings of the dissociative-programmed, fearful but arrogant mavens of the new-age movement do. I just think it's time for us to finally move away from all that pseudo-Hindu, hippie horse$#!+. Note that the caste system is a common feature of India and England, by the way: the twisted belief in 'karma' performs the same oppressive, social function as the Western belief in 'original sin' does.

Don't hold a cell phone next to your head much, please, or you might acquire some stupidity. You can cook an egg between two cell phones, so imagine what holding one next to your ear does to your brain. Cancer of the lip is the trademark of recalcitrant hillbillies, thanks to chewing tobacco, and brain cancer is turning into a trademark for phone-addicted yuppies. The little phones, which attach directly to the side of the head, make people look like androids, especially on the Europoid guys who shave their heads. Is the ear-inserted cell phone the ultimate sellout to self-annihilation?

Use the earphone/microphone plug-in or the speaker function, okay? Orgonite near the heart or by your computer and television will transmute all of the harmful energy that comes at you from larger devices, outside of your etheric field, don't worry, but the cell phone gets right into your personal energy field, so there's no effective way to completely block or transmute the close-up effects of its destructive radiation. Damage to physical tissue follows damage to the underlying energy field, of course. Before long someone will probably develop cell phones that don't hurt us, and the market will generously favor the company which exploits that opportunity, I'm sure.

Modern technology is our friend, so don't let slick, charismatic potheads convince you that the shortcut to being like the dolphins is to eschew technology, okay? A casual glance at these lovely Luddites' addictions and messed up personal lives, gives us a clear picture of who they really are.

Cars and flying machines have enabled so many people to travel freely, throughout the world, that they might represent an intermediate phase toward the end of national boundaries, as barriers to trade and international cooperation; also the end of cultural isolation. How can that be a bad thing, especially now, when the US alleged government, at least, is doing its best to restrict travel and to induce xenophobia within our borders? Sure, many people come here to seek economic freedom, but some of us who have been declared enemies of the State are also leaving to seek political freedom and to continue breathing.

What if Spanish becomes America's primary language? Wasn't that Columbus' second language, when he came here and requisitioned North and South America from countless millions of people, who didn't speak either English, Italian or Spanish? At least the Mexicans, Southeast Asians and Middle Easterners who come here, unlike the European colonists and carpetbaggers, aren't committing genocide against us, or driving us off into the deserts and mountains and the only thing they seek is employment and a comfortable life. Their culture is a lot richer than ours is, by the way, and considerably less schizophrenic so who really stands to gain from their presence, if not us?

Ever true to the schizoid nature of the propaganda organ of this government's corporate sponsor, (The What To Think Network, like all corporations in America, is a subsidiary of London's Federal Reserve Corporation), 'globalism' is also being vigorously promoted, though their version represents global tyranny, 'population reduction' in Africa, and 'political correctness' conformity, not unity.

The failed paradigm for globalism is the United Nations—an entirely unelected, would-be global governing body, sometime administered by a Waffen SS officer. Its charter was written by an American communist, who was later imprisoned for treason. The SS, which thrives and is headquartered in Switzerland these days, was never directly connected to Hitler's regime, by the way. It's an ancient secret order, rather, that was affiliated with National-Socialism's German version for just a little while.

To be a member of the SS in the late 1930s, when Waldheim was first sporting the black uniform with silver lightning bolts on the lapels, one had to murder an innocent, (usually a Jew or Gypsy in those days), in cold blood; it was a very exclusive club and nobody was drafted into it. Sicherheitzdienst Schutzstaffl, whose eagle logo has lately been adopted by the Homeland Security Abomination, is the organization that's most closely identified with the Vril Society and is perhaps that old organization's operational arm, these days, especially in the Middle East, in conjunction with Mossad and the Jesuits. At least that's what some of the psychics are seeing on our intel-gathering expeditions and it's also what is indicated in several reputable research journalists' books.

"THE HITLER BOOK", a compilation of the works of some reputable, modern German research journalists, provides documented evidence of the present activities of the SS. D Bradley told us, from personal experience, that the higher up within the Illuminati one goes, the more German is heard spoken. The Vril ritual language sounds more Slavic, though, and is called, 'Old German.' Really, really old, I expect—perhaps harking back to the day when Tibet was close to sea level, and was inhabited by fair-haired Europoids.

The US government treatied away all of its national sovereignty to the UN, in stages since 1939, though no mention of that was ever made by the What To Think Network or by your history professors. I bet you believe that the UN was started after WWII, as a gracious gesture by peace lovers. In fact, the 1939 charter, which was drawn up on a British yacht, offshore near the US, delineated the national boundaries of all the European countries that came to be adopted at the close of WWII—they were quite different from the 1939 boundaries. The reason we know about tha, is that an intrepid American newspaper publisher did an expose on that, in 1939, shortly before the start of WWII in Europe. He even listed all of the signatories and challenged the US Government to tell about how these 'representatives' were actually chosen. The transfer of US sovereignty to that lawless agency, is well documented in several books, including the US Congressional Record.

The Lucifer Trust, operated by Theosophists under the direction of Alice Bailey, immediately began publishing all of the UN literature, around the time that Rockefeller donated the land for the UN Headquarters, in New York after WWII. I doubt anyone would regard the Lucifer Trust as a service-oriented organization. After some exposure, they quietly changed the name to Lucis Trust, which still publishes all of the UN's official literature. You and I may not believe in infantilism, such as the existence of devils, but the upper level Theosophical hierarchy clearly does, and that fact needs to be reckoned with. Devil worship is always a cover for criminal behavior and criminal behavior is the hallmark of this parasitic world order.

Contemporary, terrifying, national legislation in America reads like dark-future science fiction, that would have staggered George Orwell's imagination, as I mentioned before. If most of us weren't armed, the current Bush Senior-managed regime would probably try to enforce that deadly legislation, with the assistance of millions of unseen Homeland Security Abominations, who are led, as you well know, by the former directors of the KGB and the East German Stasi. When these millions of secret police start putting on their black, military uniforms it's time for you to put that ammo clip in your pistol, partner, and keep it under your pillow, because a very short civil war is imminent then, which you're going to win. Aim for the face or groin because they'll all be wearing body armor, when they arrive.

I wonder how many of these waiting federal assassins even speak English. Should we all learn Russian now, in order to be law-abiding PJ citizens?

The fifty thousand of us who are already on the official enemy list will be taken out first, probably at 3AM, 'the day of,' and with silenced weapons, which is why I'm leaving the US shortly. For now, the federal government still have to content themselves with incremental steps toward that goal, and by now those steps are on thin, breaking ice, thank God.

As you probably noticed, the political conditions in the US are almost identical to the political conditions in Germany in 1934, right before the National-Socialists started disappearing dissidents. Hitler had been successful at schmoozing the citizenry into really handing in all their weapons, though, and this National-Socialist regime utterly failed to do so. Now, this fascist regime has to $#!+ or get off the pot, after several years of saber-rattling against those of us whom they are calling 'terrorists.' It was so much easier for Hitler, because he got everyone to be terrified of commies and he didn't have to contend with the internet and rising consciousness.

An example of rising consciousness is that the American people roundly rejected nationalized health insurance and a national ID card, fifteen years ago. Nationalized healthcare would have taxed us all into poverty, exponentially built up the enforcing power of the IRS, of course, and largely eliminated the strong middle class in America, of which I'm pleased to say that Carol and I are members. They secretly got their 'National ID card' wish, by turning all state-issued IDs into mandatory national ID cards and by requiring the use of the slave number for all commercial travel and bank transactions by now, even check cashing. That was done in incremental steps and was never announced publicly. For years, I was able to get driver licenses without a slave number, but five years ago they closed that little window of freedom, too.

In order to cash checks and money orders, I eventually had to cave in and get an Idaho state identification card, last year, which means I had to resurrect the slave number that was assigned to me, which I sent back to the IRS ten years ago. As soon as I did that, I got a letter from the IRS, which I threw away, of course. Before, my native American tribal ID card sufficed. I can still get into Canada and back in the US, with only that ID card, by the way. One of the strange benefits of Political Correctness is that even the federal cops don't want to be pegged as racist—except the beer-swilling BATF thugs, perhaps, who still hang innocent black men at their picnics, whenever they can get away with it. Somebody has to be the bottom of the federal police sewage barrel, I guess, and these guys made an indelible name for their agency, when they blew up the Murrah Federal Building, then killed a bunch of white women and children, in Texas a year later.

Now that I've indicated that I'm an Indian are you going to take me more seriously? Too bad I'm not black, I guess. Kizira, my witch doctor buddy in Uganda, told me, 'You have the soul of an African,' at least. I know I tweak just about every Europoid schizophrenic personal paradigm—and a few others—these days but, as I see it, the alternative for us, if we don't seek a broader worldview, is genocide, enabled by the old, persistent, masochistic 'victim' programming, in the deranged West and East: the bastions of materialistic civilization. It really is up to people of color to guide the rest of humanity out of this self-perpetuating, dispirited, materialistic cultural miasma.

By now Congress, the executive and the judicial branches have committed so much treason, (including the Supreme Court's bold appointment of a US President, back in 2000), that it would be a kindness to all humanity, if Washington, DC, was cordoned off by the Virginia and Maryland National Guards and all federal government representatives living outside of that little, diamond-shaped territory, were simply arrested by Sheriffs to stand trial as accessories after the fact to murderous treason, at least. That would automatically dissolve the Federal Reserve Corporation, of course, so all of our courts would have to revert to the rule of law—how can that be bad? Aren't you tired of being exploited by a London-owned corporation?

How will you feel, if our intel is correct that China purchased that paper asset, hence owns the US Government, now?

State governments, after the closing of Washington, DC, would then be held accountable for actually governing the states, instead of prospering from the 'lobbyists' generous bribes, in exchange for doing the Fed's bidding, by castrating their respective state governments.

If we're not mistaken, China called in its markers, in the fall of 2005, and now owns the Federal Reserve Corporation, including the imaginary debt that has enslaved this treasonous federal government. That's sort of like buying the Brooklyn Bridge from a con artist in New York and you'd think the old timers in Beijing would know better, since they're allegedly drawing on many thousands of years of cultural experience.

We never needed that corporation, after all, and China can't come over here and force us to obey them. The scores of millions of Joe Sixpacks, would draw their guns and start blazing away at the People's Army, if they tried it, thanks to the countless thousands of newly-malfunctioning death towers. That is, we'd defeat them, after taking a few days to defeat and round up those millions of Homeland Security Abominations. America would only be one of the fronts. China and Russia would also have to invade and conquer the rest of the planet, at the same time, in order to gain control of humanity and there are thousands of people in other countries who are also disabling those new death towers with orgonite, so the bad guys are going to fail, no matter what.

Without the death towers, they'd have to have a billion men to enforce martial law, as Russia found out the hard way, in Chechnya. The chemtrail program failed four years ago, before they could initiate phase two: genocide—the kill shot. Thousands of orgonite cloudbusters defeated the chemtrail program then. See how the chemtrails and death towers were supposed to have enabled genocide and martial law? Without genocide, martial law is unworkable, even if all of the death towers were online.

The AIDS genocide agenda clearly failed in Africa, thank God. The death towers are as prolific there as they are here, by the way. I know that, because we busted most of them in Uganda in November and December 2003. We also saw plenty of them in Namibia, when we visited there, during the time the death towers were still being erected worldwide, in December 2001. A fellow, whom we sponsored to go do some gifting in China, saw them all over Tibet then, too. I bet there are more death towers than cell phones, in Tibet.

Have you read the US Constitution? It's a quick, easy read and not ambiguous at all. There's no lawful basis for a standing army in the US, any more than there is for one in Costa Rica. The National Guard, in the various states, is sanctioned, though—an 'on call', competent but part-time army for the protection of our borders. Our national borders aren't in Germany, Japan, Korea, Vietnam, Afghanistan, Somalia or Iraq, of course.

There are no viable foreign military threats to America, or just about any other country. Nations—most notably, Tibet and Iraq—that have been invaded and occupied by foreign armies, now are more than just threatened, of course, but when the corporate hierarchy has been displaced, the rest of the world can right those wrongs, with 'consultative will' and appropriate force, exerted by an ad hoc military campaign, as a last resort. There are plenty of trained soldiers in the world, and plenty of equipment, so maybe some of it will come in handy someday. If there's no big, centralized political power in the world there's not going to be a threat of tyranny, especially if the rest of us nip it in the bud, if it starts.

The only Muslim terrorists operating within the US are employees of the CIA and its close affiliates. That, too, is abundantly documented, even on the internet. Every terrorist organization in the world is a corporate organ, so dissolving the few corporate-sponsored tyrannical regimes, by destroying the London-based corporations, themselves, will probably quickly end this fake reign of terror. Mossad and the other terrorist groups, all get paid by the same London/European bankers, including the Fed, whose chief executive leg-breaker is Bush Senior, of course—he's been London's global wise guy, since 1980. This has become so widely known, that it's probably just a matter of waiting until the information becomes 'common sense,' which is to say that the Pajama People, themselves, will eventually become aware of it and can still remain asleep, in spite of knowing it.

That will signal the quick end of corporate rule, after all, just as corporation-sponsored, 'legal,' slavery ended in the world, by the close of the US Civil War, due to enough public outcry. Even the Indian tribes in the US—not to be outdone by Lincoln's gesture—freed their slaves then. A German, Queen Vicky, ended it in the British Empire before that, as I mentioned before.

It shouldn't be any more inconvenient to cross over into another country, than it is to cross over from Tennessee into Arkansas, really. Genuine, corporate-sponsored lawbreakers and other threats to national sovereignty, get around those border inconveniences with little or no trouble, after all. Nelson Rockefeller, for instance, used to be met by mandatory cheering crowds at the airport, in Moscow in Khrushchev's day, and he didn't even go through the customs line, for instance. During that time, Khrushchev banged his shoe on his desk in front of the UN General Assembly and told us that he was going to bury us. What a hypocrite! The Russians literally rolled out a red carpet for that mass murderer, Nelson Rockefeller, then. See how hopeless the situation was for JFK? He thought that if he only disbanded the CIA, life could be sweet for humanity.

If the horrific Patriot Acts were openly enforced, the Pajama People themselves would angrily wake up, draw their pistols, and the enforcement would end in a day, as would this cancerous federal government. We don't want that to happen suddenly, because we might not recover our material culture, if anarchy got started. The ridiculous local cops, who shave their heads, wear dark sunglasses and dress in battle gear, would be the first ones to fall, in that case, of course. No matter how brutal and intimidating these guys look, lately, they can't hold out for five minutes against an armed populace, if the cops were suddenly ordered to round up dissidents, and they most likely wouldn't even try. We've been told by someone who knows something, that in many areas where there are actual patriots, (active Constitutional militias), these thug cops all carry a change of civilian clothing in their patrol cars, 'just in case.' I didn't hear whether they carry toupees, too.

There aren't many cops like Police Chief Billy Phillips, in Tennessee, who participates in our network and is beloved by his town folk. We met him when the FBI was trying to frame him and his lieutenant for openly opposing the Patriot Act, in a regional police chiefs' meeting. Carol and I smacked the FBI guys silly, etherically, and we sent Chief Billy a bunch of orgonite and a Succor Punch. The FBI withdrew, then, but tried again a few months later. This time, Billy was pretty good at smacking them, too, and they withdrew again. That was a year ago, and he's been fine ever since. He gets invited to give talks to patriot groups about the US Constitution and the rule of law. When we defeat a lawless police agency, they never admit defeat; they just go away.

Each of the American states are nations, and the US government was set up as an intermediary and coordinating body, by these states, when they were third world nations, directly threatened by Great Britain at the height of that country's ruinous, parasitic, empire-building career. Did you know that the City of London invented the MalWart paradigm? The Chinese simply copied them, lately.

I think California is the only American state where every driver is stopped, photographed and questioned at the state line, (frontier), but if Congress gets their wish granted by the Homeland Security Abominations, every road on every state border will be a police roadblock and checkpoint. 'Sieg, heil!'

As recent precedents have shown, the feds would dearly love to involve the US military, full-scale, in domestic police work, and they have been trying it out on the populace for the past twelve years or so, mostly in urban areas where people are stupidest. This is expressly forbidden by the US Constitution, of course. They would have initiated a full-scale police state, under martial law, but they figured out that not enough of the American soldiers and marines would support that, even if our military force was large enough. At the current strength, it's likely that the entire US Military would only be able to occupy a single major US city, not the entire country, and as the Warsaw Ghetto's inhabitants demonstrated in 1943-4, even a single city might be too much to handle, in that case. So, if you feel intimidated by the What To Think Network's babbling about martial law, what is your fear based on?

A fellow told me yesterday that an elderly friend of his, who is from Europe, claimed that the SS put fluoride in the water at the concentration camps, so that they could guard the inmates with a third as many soldiers as would have been required, otherwise. I wonder how many of those SS policy makers were also dentists.

Nearly all of our under-equipped soldiers are in Iraq, and we've seen evidence that they were sent there after they made it clear that many of them were considering exercising a coup d'etat against this fake president. I wonder if they're all being set up by the US, (Chinese?), Government for wholesale slaughter, by Iraqi patriots, if the Iraqis ever decide to take their country back, after the cakewalk that was being sold to you, by the What To Think Network, as 'the liberation of Iraq.' We know that the car bombs are all being set up by US Military Police, at checkpoints, so you can bet the Iraqi people all know that, too. What would you do about that, if you were an Iraqi in Baghdad? That's only one of the many things that are being done by US soldiers, to alienate the Iraqi people. Each of those soldiers took an oath to 'defend the US Constitution from all enemies, foreign and domestic,' and to 'disobey all unlawful orders,' so you might say that they deserve whatever calamity might befall them in Iraq. I hope to God they'll come to their senses and just come home, before something awful happens to them.

Even if all of the states, (except California?), functionally seceded from the union, we'd still call ourselves Americans, I think. The only difference in our lives, in that case, would be that we are no longer taxed into poverty, no longer spied upon and intimidated, no longer conscripted to be cannon fodder in a foreign country, no longer required to get permission to travel on the highways or to carry costly insurance, no longer required to report our earnings, no longer required to relinquish our children, at birth, to a foreign corporation, no longer forced to use the national ID card to do financial transactions or get on an airplane, no longer ruled by appointed corporate bureaucrats, etcetera.

The armies of Russia and China are the largest, best equipped and most mobile, in recorded history, so what's to be done about that very real threat to world peace now? I've been thinking about that a lot and I'll throw this out as food for thought and candid, public consultation:

These armies were built up since WWII, at the behest of the same European bankers who were given control of the US government, in the 1930s.

Russian military have thoroughly infiltrated US cities, as 'the Russian Mafia,' in close cooperation with the CIA, as street-level dope pushers. No mention was ever made of 'Russian Mafia', until right after the end of communism in Russia, when it was announced that their vast occupation army in Eastern Europe would not be allowed to return home. They mostly left Eastern Europe then, so where you do suppose they went? Mongolia? France? Which nation is roomy and wealthy enough to warehouse millions of Russian soldiers and perhaps support them with the drug trade?

When all of this was happening, NPR and the rest of the What To Think Network, discussed almost nothing else, but Bonehead Clinton's dalliances. The whole country looked pretty stupid in those days, to the rest of the world, thanks to our media, and it's still remembered. A couple of years ago, for instance, Dr Kayiwa and I passed a little café on the outskirts of Kampala, Uganda, called, 'Monica Lewinsky's Joint.'

Clinton, at the behest of Bush Senior, who became so unsavory as overt president, that the corporation had to put that lovelier, charismatic pigeon in front of the camera, to distract us all from the man behind the curtain, granted the Chinese army its own container port, at Long Beach, (LA), at the same time that he gave them the Panama Canal and the US Patent Office. I bet you didn't know that Chinese troops operate, therefore essentially own, that canal. A railroad was then built from the port at Long Beach, and you can bet it wasn't for MalWart's convenience, because those goods were already coming into America, without a hitch and probably duty-free, probably even subsidized by Greasy Greenspan's largesse.

Soon after that, imported Chinese arms were openly distributed by the CIA, among Los Angeles' drug gangs—that was the federal hand caught in the fortune-cookie jar, for a change. The CIA wasn't expecting the resulting public outcry from LA's Black citizenry, and were caught with their pants down.

You probably, at least heard, about reports of Russian soldiers in America, since the time of the federally-sponsored massacre outside of Waco, Texas. Some eyewitnesses claim that Ukrainian soldiers and one of their tanks were involved in that massacre, along with those fat, leering, ubiquitous BATF baby killers.

You probably also remember that some of our national parks have been closed to the public, since that time, and that air traffic is restricted there. Maybe you even saw some of those Russian tanks and other military vehicles, painted white (UN), on flatcars back then. The Russian stuff is old news, in fact, but not the Chinese military activity in America. Carol started seeing Chinese military in underground bases in LA, when we did our first gifting there, in June 2001. That's the first time she saw Chinese, among the institutional predators in America, and I wondered if it wasn't the exception. Now, every time the psychics in our informal blasting group look for culprits, they're seeing Chinese military guys at the top of the Homeland Security Abomination teams, who are regularly assaulting us all electronically, psychically and with poison.

Take heart, though! You may have seen that these heinous, corporate people, whose wishes for global tyranny and genocide are not being granted in America, are also meeting new obstacles to tyranny, even in Russia and China, apparently.

You won't find anyone more genuinely optimistic than I am, and I'm one of the people these tyrants hate the most, as you might imagine. The initial shock of noticing the level of tyranny that has been established is a necessary step toward gaining confidence in the future, because the enemy is now clearly identified. Before, they had scammed most of us into fighting each other across ideological, racial, cultural, political and other meaningless divides.

If you want to acquire confidence and faith like mine, start making and distributing orgonite and if, in the process, the Homeland Security Abominations show their ugly mugs to you, learn how to throw energy at these predators and take the initiative away from them, as so many of us are now doing. You can rightly figure that one of us easily overcomes a hundred of them, short of resorting to a shooting match. God willing, it won't come down to shooting, if we do our healing jobs right.

The US Congress and the fake president, have essentially established martial law in America, and some Chicken Littles, mostly on the internet, actually claim that this will be enforced—they don't mention 'by whom,' because an answer like that would require some critical thinking. Who could enforce martial law, among an armed population of almost three hundred million? I feel very encouraged, by a recent chain of events, that the ruinous, European/Fed-sponsored Chinese and Russian regimes are about to follow this treasonous American one into oblivion, in coming days. When the fall of tyranny starts in those countries, it will accelerate quickly. Look how close it came to that in China, right before those kids were slaughtered in Tienanmen Square, for instance. You can bet that the Chinese people didn't forget about it, because most of them trusted their government before that, and that trust was irreparably destroyed in a day.

I think we can actually benefit from the disbanding of those vast armies, by giving each soldier a few hundred acres of Sahara or Gobi Desert, a tractor, a prefab house, a well, an orgonite cloudbuster and plenty of food and cash to help them get started. The gold that used to belong to the US Treasury, which we apparently never needed, can easily finance all of that, after the fall of the Beijing and Kremlin parasite dynasties.

That might seem frivolous to you, if you believe that only governments can undertake the reversal of deserts, but in fact, Georg Ritschl is doing that, systematically and almost single-handedly, in the Kalahari, these days. A hodgepodge of gifters, including Carol and I, have been reversing the American Southwest's desert, in recent years, including Death Valley, as I mentioned before. Just because the What To Think Network ignores it, doesn't mean that I'm not right. I think they screwed up, when they put Laura Bush in now-lush Death Valley, for a photo op last year. The valley was as dead as ground zero, on Tuamotu Atoll, when we put our first cloudbuster there, four years ago.

When you consider the amount of potentially arable land there is on this planet, even the present population growth rate might be insufficient to get these areas settled and green again, within a generation. By the time that's accomplished, we can terraform and colonize other planets, of course, if we start to feel crowded.

I personally believe Al Bielek's claim, that the world order has had interstellar craft, since the 1980s, and that this new tech takes advantage of 'worm holes'—interstellar shortcuts in time and space. Many of us assume that our own planet was seeded that way, perhaps several times, by several sentient species, including human, reptilian and cetacean ones. If that notion disturbs your belief in God, please review some of the history of the cynical manipulations of your religion's more disgusting clergymen. What I'm talking about lends even more credence to the notion of a Maker, than a Michelangelo painting of God's bellybutton ever could.

Others' documented claims about interplanetary craft, (which corroborate Al Bielek's claims), that have been produced by National-Socialist corporations since WWII, are convincing alternatives to NASA's cynical, costly, Apollo scam. There are many, strong evidences that those moon landings were fake, but my favorite is that, in order to get through the Van Allen radiation belt without getting cooked, one has to have extremely heavy and bulky lead shielding, or to create gravity/EMR stasis, in and around one's craft, which is what the little saucers, (the ones that were actually used to create bases on the planets), can do. Otherwise, the only way for living organisms, besides just cockroaches, to come and go from other worlds is from above the poles. The van Allen belt is a toroid, open in the north and south. Carol and I have seen those silver saucers from several angles, even up close, so I know they exist, not that my eyewitness accounts can do much to convince you, of course. Keep looking up and you're bound to see one, especially if you're an avid gifter. We usually saw them on critical gifting sorties.

The UN promotes 'population reduction' openly in Africa, which is a polite, newspeak term for 'genocide.' The World Health Organization, which most Africans despise, is the agency that spreads AIDS, throughout that continent. Half of Africa is uninhabited desert. Much of that was farmland until fairly recently, when weather warfare was perfected and deployed in Africa. Our East African cohorts, to the north of Georg's Southern Africa stomping ground, routinely end severe droughts these days, by simply distributing orgonite to villagers, I'm happy to report. If Carol and I hadn't repeatedly witnessed how fast and profoundly, even a little bit of deployed orgonite affects the atmosphere there, we might have a hard time believing it, ourselves, so you sure have my sympathy, if this claim seems unreasonable to you.

Carol and I knew a year ago that the we'd probably be interacting with dolphins and whales at sea by now, and it took us awhile to arrange our affairs to be able to move back to South Florida in September 2005, to get a boat and start gifting the Atlantic, in this vicinity. Progress has been quite slow, and we've spent about forty gallons of resin, making gifts to toss into the local seas, in the past six months.

A whole lot of obstacles have been put in our way, before and after our arrival in Jupiter, Florida, including box surveillance by several federal agencies and even the local cops and Sheriff deputies. But that all just feels like confirmation and encouragement to us, more than just an annoyance. Killing our dog was over the top, even for them, of course.

We got a kick out of one Homeland Security Abomination, who pretended to pick up dog doodoo for money, behind the little grooming facility next door, in order to get close enough to attempt to get our dog's confidence. I don't think their tactics get much lower than that. Fortunately, our pup was smarter than that cop and remained suspicious. I don't think he's the one who killed her, because he stopped showing up, when he showed his hand that day, several weeks before she was killed. I had a hard time suppressing my mirth, when he came into our yard, uninvited, and tried to give her a big dog biscuit. Another Homeland Security Abomination tried that, in Key West, and she nearly bit his hand. Dogs are good at discernment.

By now, we've tossed enough energy at the more enthusiastic peekers and burglars that they remain discretely out of view. Every time we move to a new area, we have to cycle through all the fed cops and gung-ho, shaved-head, local fascist cops, who get in our faces, tag along on the roads and peek through our windows. You'd think that our reputation would precede us, but I guess people who choose to work in fascist police agencies aren't generally quick studies, nor are they prudent. We caused several Fed cop hideouts in our previous town to shut down, with gifting and frequent blasting and, for a while, we were going through a new SAIC every week or so. That's 'special agent in charge', the boss Fed in a locale. I don't know how much hatred has been directed at you in your life, but the other side of the 'making friends' coin is 'making enemies,' if you're in the healing trade in America these days. The hatred we get from blasted Feds is palpable, and I once rattled DB pretty good, when we were gifting together in Hollywood a few years ago. Since he's telepathic, he could tell that one of the CIA tagalongs I was taunting came close to shooting me, out of frustration. That Fed looked pretty mad.

An example of bizarre misogyny/masochism is the way some of the more enthusiastic fascist cops tailgate Carol, through town and on the highway, even though she hurts them with tossed energy in the moment, a lot more severely than I do. When I drive around, there are no cops to be seen, even though they know I don't have a driver license, and the feds who watch the house from several directions, through video cameras, no doubt, would like for them to lock me up on that account. Carol has the impression that if they locked me up, it would cause more trouble for them than they're willing to deal with, but I don't wish to test her theory; not that I couldn't use a vacation to catch up on reading and do some fasting. I'm awfully busy these days.

The Feds arranged to have me locked up in Las Vegas, a couple of years ago, and I think they assumed there was an outstanding arrest warrant for me in Washington, State. There was no warrant, after all, so they had to let me go. A year earlier, they came into the house and injected a pretty strong poison, perhaps radioactive, into my left forearm and marked up my chest with a poison needle, but the zapper and orgonite took care of that. I only spent a couple days in bed and that hasn't happened, since. I've got a nifty scar on my arm, though. I think they were especially frustrated, then, because Carol and I had just managed to get DB's film, "CHEMTRAILS: CLOUDS OF DEATH", on the internet.

We first became aware of overt, federal surveillance, right after we invented and set up the orgonite cloudbuster, in Florida in March 2001. We feel a need to complete what we came here five and a half years ago to accomplish, which meant getting a seaworthy boat, in the short term, that can get us comfortably, quickly and safely to the Bahamas and Yucatan and back.

This book might be on the market, before we get into the water with the local dolphins, or get over to the Tongue of the Ocean, (south of Andros Island, Bahamas), and the northeast point of the Yucatan Peninsula. But I feel that it's important to get something about our history and work together, into the printed record for now, due to some public and hidden moves that are being made by the sewer rat agencies to assassinate our characters.

Our accomplishments to date probably warrant a published personal account by now, at least, even if the books are only available from Amazon. I think books can be sort of like bullets in an etheric war, in terms of defending against character assaults and ending the insidious, subsurface tyranny of walking, corporate parasites, who are employed to savage the reputations of all of the genuine pioneers in the world, or, failing that, to overwhelm legitimate information sources with a flood of disinformation sites, glitzy books, personality cults and What To Think Network programming.

I wish to lift the lid on these activities, because parasites always dread exposure, as well they ought to. They also dread our active ability to think critically, of course.

Typically, when any individual sets out on a path of service and inspires others to do the same, the hidden but pervasive, (at least in the West), parasitic hierarchies will do everything possible to prevent that person from fulfilling his or her destiny, because the genuine empowerment of one human is the empowerment of all of us, especially now. Anyone in Russia, China or some of London's 'Islamic Republics', who overtly tries to heal the environment and/or liberate the huddled masses, for now, still just gets Tienanmen Squared, of course.

Short of efficiently suppressing us, those agencies will cut off their own noses to spite their faces, if necessary, in their constant effort to discredit genuine pioneers like Carol, me, and a few other notable people in this movement, whom I'll continue to tell you about, and who have something worthwhile and new to offer. Last summer we witnessed a weird spectacle within the internet orgonite network, when a mole, apparently a KGB asset, ousted several very resourceful and patient CIA and MI6 moles, on Etheric Warriors and on another board, in an effort to divert attention from her own questionable affiliations and ethics. She also disabled most of the death matrix, in and around a major US City, in the process of gaining our confidence, earlier.

Carol was personally stung by this person, because she had become a confidant and close companion. It's a lot harder for Carol to make friends than it is for me, so betrayals tend to hurt her more deeply. I think that if I were psychic it would be harder for me to make friends, too, because it would be harder for me to ignore others' flaws.

The swelling cadre of free energy device inventors, have been successfully suppressed by these corporate agencies for over a century, unfortunately, but maybe some of them will eventually adopt our methods of self protection, then maybe they can start marketing their wares with impunity, as Carol and I have been doing for years, with our cancer-curing zappers. A century ago there were only two or three free energy device inventors, including Tesla. Now, there are dozens; perhaps hundreds. I bet the assassin agencies are chasing their tails, surveilling this many inventors.

Most of them are too proprietary to work well with others, though, unfortunately, and it's unfortunate, I think, that so many of them invent exclusive, new words to describe what they're all talking about: orgone/ether/chi, etcetera. Implosion is the energy basis of all free energy tech, the way electricity and explosion is the energy basis for all the tech in the receding paradigm.

I personally knew Wilhelm Muller, the inventor of the magnet-powered motor/generator, and I played with his quiet, 400-horsepower prototype a few times, and learned everything I was capable of grasping from his exhaustive explanations, diagrams, endorsements and demonstrations. At the time he was murdered a couple of years ago, he was in a position to get a small, hundred-horsepower version mass-produced, in a Vancouver factory, for around $1,300 apiece, retail. I want to eventually put one, each, of those, in an ocean boat and airplane or helicopter.

I put a homemade Joe Cell on our truck, and it came a hair's breadth from changing the motor over from running on explosion to running on implosion, in August 2002. The center bolt of the cell collapsed, from the violent shaking of the transition stage, before I could advance the distributor and stop the shaking. Orgonite seems to be the key to keeping a Joe Cell running, in areas where deadly orgone radiation is present.

With hindsight, it's possible that The Operators prevented me from achieving success, so that I wouldn't be shot by frustrated Federal cops, because you can probably guess that I'd have broadcast my success immediately.

Look at how many healers have been murdered or imprisoned, during the generations-old Federal War on Healers in the USA alone, before we came along. At least the internet threw a wrench into that war machine, thanks to it's immediacy. When I put my zapper business on the web in 1998, I knew I'd come home, and the business has grown steadily, by referral, ever since, so that now I'm selling more zappers than anyone else, including Hulda Clark. Carol and I have been routinely curing cancer and AIDS, for years and telling about it on the internet and even in popular international radio show interviews, a couple of times.

We eventually got zappers into mainstream awareness in East Africa, thanks to Dr Rushidie Kayiwa's networking among the movers and shakers, and Nathan Kagina's persistent, successful, publicity efforts. Nathan's AIDS cure testimonial and his before/after pictures are on our site, www.worldwithoutparasites.com.

We're prudent enough to understand that it might become necessary for us to duck into Central America or Africa, on short notice, to continue our public career, though, if the medical mafia and/or Homeland Security Abomination turns up the heat under us, in the US. Somehow, I believe we'll continue to be relatively safe from those bloodthirsty terrorists, though. Let's see if publishing a book puts us into a more vulnerable position. I'm pretty sure that if I was featured on a popular TV show, I'd have a hollow point 9mm bullet in my brain, shortly after the end of the broadcast, because TV is the occult/corporate order's sanctum sanctorum. The internet is this network's sanctum sanctorum, and the murderous old corporate farts are still too blind to see the internet's raw, unfolding potential. I think books are somewhere in the middle.

I started using pennies as electrodes on my zappers many years ago, because they're a delicious political statement. Much of the medical mafia's proxy sabotage effort in the alternative healing forums, (postings by a coordinated, professional crew of clever sociopaths and some support technicians, mainly based in Canada and apparently sponsored by MI6), has been directed at our use of pennies, (especially the head penny, which has the image of that otherwise-useless Whore of Babylon on it—QEII), but fortunately, our happy and loyal customers stick up for us every time, in those forums by countering those sociopaths' aggressive, anti-scientific assaults with effective, rational arguments and personal testimonials, so we end up selling even more zappers.

The paid sociopaths have pretty much thrown in the towel by now, but a few years ago they were able to cut our sales in half in a short time. That happened soon after we put the information about disabling the new death towers on the internet. They knew that our zapper business was financing our gifting efforts. We didn't slow down the gifting, though, and we generated ten thousand dollars of credit card debt, to keep our momentum. Fortunately, within a year our sales were back up again, and we paid off the credit cards.

An attempted double whammy happened during the time they cut our sales: Jeff Rense dedicated several of his national radio show broadcasts to discrediting orgonite, and warning people not to make orgonite cloudbusters. The attempt backfired and doubled our network's numbers within a month. Rense abruptly stopped mentioning all of this after that. I don't think he even has critical material on his site about orgonite any more.

The attacks on our zapper business in alternative medicine forums mostly stopped a couple of years ago, and the other side had to resort to even less effective methods, such as last summer's E-Bay scam, which was also easily exposed and neutralized, thanks to David Huddleston's timely assistance. David sells our Terminators and Harmonic Protectors on E-Bay with an ad, which he updates a lot. The would-be business assassins provided some personal fun, actually, when I took apart one of their 'Inexpensive, Don Croft Style Terminator Zappers,' which was a camera battery connected to two penny electrodes, which were glued to a little black box full of silicone and bits of broken, colored glass. I sent the contents to David, who photographed it all and put the photo in his E-Bay ad with an appropriate caption. The ensuing debacle reminded me of the verse in the Qur'an, 'Behold the confusion that hath befallen the tribes of the defeated.' That scam, too, was operated out of Canada, apparently by MI6. They used Canadian pennies on their fake zappers but only the tail sides.

Those fellows went down pretty hard then, which was gratifying to me and woke a few PJ folks up to the reality of their employers' old, international 'War on Healers.' I think they were mainly aiming at our gifting work. The bad guys are a hundred times madder about orgonite than they are about zappers.

It's the overt measures like this, and like the chemtrails and death towers, that ultimately lend credence to folks like me, who contend daily with the mass-hypnosis programming that's based on non-conspiracy theory. This is the first time in history that tyranny is widespread and overt, so more people are waking up now, than otherwise might have. Everyone who finally looks up and sees chemtrail remnants, death towers, battle-dressed small town cops and Airport Gestapo, is enraged, for example.

There is no enduring empowerment in this life, except through the genuinely humble pursuit of selfless service. Who is there, now, to teach us this simple lesson better than the dolphins and whales? Carol feels strongly that the next phase of humanity's salvation rests with the waiting cetaceans' demonstrations and instruction for us all. They love to receive orgonite, so everybody has a handy, potential interface with these wonderful, mysterious and vital creatures. Maybe the distribution of orgonite to the cetaceans for healing our oceans is another bridge to world unity and prosperity, in a way that's roughly similar to how the internet has been facilitating it, for the past decade or so.

Reality is made up of more enfolding layers than we can possibly track, but what I know for sure, is that the organized exploiters in this world are ultimately just finger puppets for The Operators, so when these people manage to throw an obstacle in our path, on one level or another, it's probably only because the Good Guys find it prudent to slow us down or enable us to strengthen our resolve. Only genuine faith carries us forward, when things get rough; ideology is worth nothing then.

Ideologue: an impractical idealist.

If you effectively oppose the established, but waning order, as many of us are now doing, they're going to throw everything they can at you to get you off balance, or to terrify you back into inactivity, because this old war against humanity is mainly psychological, rather than physical. They'd rather disempower and exploit you, than shoot you. Their rule is based on our voluntary, collective acceptance of their programmed, delusional worldview. Take this a step at a time, and get your confirmations along the way, so that your faith can sustain you, when you're under psychological assault. At some point you'll see their attacks as signals of your own effectiveness. After that you'll welcome the assaults, even when they take the form of implanted, inappropriate thoughts and emotions or even administered poison. After you no longer dread them, the assaults start to decrease, because even the sewer rats don't like to waste their energy, when the results are counterproductive. Sharing your experiences with other open-minded people is also a good way to consolidate your progress.

If you smoke any pot at all, you're not going to get past that barrier. Pot, even if it's only smoked once a month, rips and maintains holes in one's energy matrix, and the holes are open invitations for interference and manipulation by the secret police agencies' psychic predators and their associated cultists, including the What To Think Network.

Ideologies divide people into prejudiced groups. If you love a particular ideology, make sure that you love humanity more than that. Only fakers say that they love individual humans, but hate humanity, or vice versa. Hating our fellows is really just self-loathing and that's an illness to which we're all susceptible. Love, courage, accountability and acceptance are a great deal less contagious, unfortunately, but exemplifying and promoting all of that is always worth the effort.

Life's a lot easier, when we're able to love ourselves and everyone else for the sake of our Maker. Any lesser motivation guarantees a harder life, on account of having less genuine detachment. Programmed dissociation isn't detachment, by the way; it's a mental sickness. In order to get properly detached, we need to get a clue about why God allows suffering and exploitation, and that takes some reflection, of course. Mumbling repetitive prayers won't give us that understanding, and nobody else can settle that question for us. It may be that we can only get answers like that, through prayer and reflection.

Westerners are so jaded by the cults by now, that mentioning 'God' or His Messengers more than once or twice, causes western democratic liberals, (the cynical products of the hippie generation), to lose interest in anything else you have to say, so I've opted to dwell on what Carol and I consider to be a more discernible—but still sacred—unseen hierarchy who are looking after us and are guiding us toward a long-promised, eventual Golden Age. All creatures serve our Maker's will, of course, but our choice in any moment, is whether to serve that divine agenda voluntarily or in rebellion. When we lawfully oppose tyranny we're not technically rebels; the tyrants are rather rebelling against natural law, by exploiting us all.

I've mentioned Carol's Wiccan practices, but I hasten to add that she doesn't subscribe to any organizations which call themselves, 'Wiccan,' and that she knows, as well as anyone, that these groups are mostly corrupted, as any other organization is, these days. This form of belief is anti-organizational, so formal Wicca organizations are mainly made up of angry, dominant, bull dykes and new-age sycophants, of course.

Carol gets attacked publicly for being a witch, and it hurts her feelings, when that happens. If I weren't relatively impervious to darts like that, maybe there wouldn't be a gifting network, because I feel like a damn pincushion on some days, when the saboteurs gang up on me. I often get poisonous correspondence, and I am told of occasional, public, character attacks, usually directed at me by complete strangers. I won't even discuss my own religion, the Baha'i Faith, unless

I'm specifically asked about it, because people are so culted out by the fake stuff by now, and I make it clear that overt proselytizing is forbidden on Etheric Warriors. Defending my Faith would be a fun sport, though, if anyone felt inclined to attack me because of my personal beliefs.

Having experienced the demolition of my ego through emotional trauma, twelve years ago, also a couple of fire-walks later on, those personal attacks mostly feel superficial to me. I hope you can achieve a functional level of detachment, without having to experience emotional trauma, first. The fire-walks were fun, but it takes a few days to get the soot stains cleared away from one's feet. Some folks wear socks and I'm going to try that next time.

MLK, Jr, is absolutely correct, that we can only be judged according to the contents of our characters. A couple of years ago, there was a concerted campaign by former associates, to discredit Carol and I, on account of her personal beliefs, and that failed, I'm happy to say, but left Carol a little more withdrawn for a while. I think that was mainly aimed at discrediting her Harmonic Protectors, which are superior, inexpensive devices for transmuting ambient deadly orgone radiation, strengthening and healing one's energy field, and blocking most energy assaults—even strong, directed, microwave beams from satellite weapon platforms, (rare, but it's happened).

Some of us take personal betrayal harder than others. I never mention any of them publicly after the fact; I simply kick them out of our forum and move on. They've already discredited themselves in the process of attacking us, and are rarely heard from, because nobody in the active network takes them seriously after they put their cards on the table.

A good, well-earned reputation is priceless, but is too-easily destroyed with slander. I think it's kinder to shoot someone, than to destroy his reputation with backbiting and lies, because murder is a temporary effect, but slander keeps on destroying, sometimes for generations.

Magic (noun):

1) The use of means, (as charms or spells), believed to have supernatural power over natural forces
2) An extraordinary power or influence seemingly from a supernatural force
3) Sleight of hand

'Magic', in the case of what Carol does, and also what characterizes our frequent, spontaneous group intel-gathering, healing and predator-blasting efforts, is simply performing work from a distance or over time. When a group effectively focuses attention and energy on anything, synergy is developed and that's also a magical process.

A popular conception is that anyone who engages in practicing magic becomes corrupted, but we're seeing the opposite happening to people who get involved with distributing orgonite and taking down occult/corporate hierarchies' leadership: we develop more humility, appreciation for others and genuine detachment, by doing this work. You won't meet a more self-effacing person than Carol, and the better she gets at performing beneficial magic, the humbler she gets.

If one feels inclined to include rituals to do beneficial work, who is to say whether any of that is improper, as long as nobody gets harmed in the process? I don't do spells but I also don't have Carol's talent or vision. I know Carol's trustworthy, so I simply assume that her efforts are useful for us both. I often see consistent, beneficial effects of her spells, all of which are done to sweeten her, our and others' lives. She won't help anyone who doesn't ask for it, and she won't interfere with anyone who isn't directly interfering with us, our friends or our families—the exception being that, of course, we all overtly interfere with the leadership of the occult/corporate hierarchies, who are exploiting all of humanity. Everyone ought to be interfering with them right now, by any lawful means. If we don't take tyranny personally, we're not paying close enough attention and, as a Greek philosopher said, 'If we won't involve ourselves in politics, we'll be ruled by inferior men.'

She doesn't promote any ideology, nor do genuine Wiccans in general even agree on very much, except the practical applications of magic. Orgonite is magical material, because orgone, (life force), in its various forms, has always been the energy matrix for the dynamics of all kinds of magic, and orgonite perpetually changes ambient, bad energy into good energy. I think it's fitting that Carol has equally shared the achievement of initiating this revolutionary, global healing movement with me, because empowered, effective women have been the primary targets of this patriarchal, parasitic order for millennia and maybe Carol's continuing survival and success signals the end of their ancient 'misogyny imperative.' More women were slain by the Inquisition, than Jews and Gypsies were murdered by the SS and, in real terms, the SS were far more humane about it. Wouldn't you rather go to sleep in a gas chamber, than be burned alive or tortured to death?

Misogyny is 'hatred of women.'

Did you know that women are generally better psychics than men are? People in the genuine, indigenous, traditional magic orders know that women are primarily mental, and men are primarily emotional. It's not hard to tell, of course, but a very big part of our programmed schizophrenia induces us to assume the opposite. In Africa, most of the good witch doctors are women and most of the bad ones are men.

The only time she came close to crossing the line, was when she induced her previous husband to stop drinking, (she used an amethyst), after their daughter was born, almost twenty years ago. He remained sober until after Carol married me, and she had undone the spell, when Jenny moved out of his home and into ours. In that case, Carol had performed the spell for the child's safety, because the dad was an alcoholic. He certainly wasn't harmed in the process, but she still feels some remorse for even doing that much, to interfere with another's free will.

She never feels remorse for what we do in the regular group intel/blasting sessions, because that's all a healing exercise, performed on the body politic of humanity. Institutional oppressors are like terminal cancer, after all, because their ultimate aim is the death of the host. Curing cancer in the body is an aggressive act, so why would curing a cancer in the body politic of humanity be any different? If you and I won't do it, who will? The Vatican? Sai Baba? Pat Robertson?

If you get involved with any of these informal energy-tossing groups, remember that energy needs to be thrown at specific targets, usually human, in order to achieve a lasting cumulative effect. The talented but misguided people, who attack 'devils' or 'heal the whole world' are just pi$$ing into the etheric wind, so any stimulation they might get from the activity is mostly what bounces back at them, from relatively impervious, vague targets. I think that in most cases, these folks are delusional and susceptible to the suggestions of CIA and other agencies' psychics, who monitor all of our activities in the ethers and manipulate us, whenever they can. It certainly serves their agenda to induce these people, who are usually arrogant, sociopathic, new-agers—go figure—to tilt at windmills, rather than at their own programmers.

The genuine witches and witch doctors I've known, here and abroad, would rather cut their own throats than harm anyone. The dirty ones I've encountered take pleasure in exploiting others. There doesn't seem to be a middle ground, once someone has committed to doing magic.

Gifting is strong magic which anyone can perform, and it's always beneficial—unless you aim very badly, while tossing orgonite out the car window, during a gifting sortie, and hit a pedestrian, or damage personal property. Gifting from a convertible adds another dimension to the fun. We did Hollywood, Burbank and Beverly Hills, that way with D Bradley.

Throwing concentrated life force from our hearts at predators, is always done according to the requirement of natural law; the outcome won't necessarily conform to our own wishes, but it will always be appropriate.

Zappers operate the same way, within the human body, for that matter. Micro-current, which is a more mundane expression of life force, simply ionizes the internal environment. Weak or ailing tissue is restored by micro-current; at the same time, pathogens and poisons are neutralized, (oxidized), by it. Colloidal silver and ozone are simply less efficient ways to deliver ions into the body's fluids. There are no mysterious, hidden processes in play with simple ionization, but adding subtle energy components to zappers induces some pretty remarkable processes. In that case, a zapper is more than just a curative device; it induces the body to heal itself and provides specific energy for that process.

All we do toward curing an illness, and healing the body with a zapper that has subtle energy components, is, switch it on and put the zapper's electrodes against the skin in a sock, or under a bra strap. We sleep with it on our sole or palm.

We mostly aren't aware of what's happening, until sickness symptoms suddenly disappear. We might expect a zapper to make us more psychic or turn us into athletes, based on isolated testimonials, but if an outcome isn't appropriate for us, it's not going to happen, nor will I ever promise specific results or sensationalize the process.

Psychics in western cultures have usually gotten the $#!+ end of the stick in their own communities, by being alternately scorned and idolized, depending on the momentary, usually manipulated whims of the Pajama People. I think clergy and masons have been behind most of the persecutions of female psychics and healers, during the past five centuries, though of course most clergy are also masons or equivalents, so it might be a moot distinction.

The joke on all misogynistic secret societies and religious orders is that they need females to conduct their higher, (sic), hidden, temple rituals, because men just aren't capable of it. That's a confirmation of the notion that women are primarily mental, so are generally more focused than men are. The perversion of the Diana cult, to which Princess Diana was apparently sacrificed underground in Paris, is an outcome of this queer, schizoid dynamic.

The LA freeway system is actually a traced outline of a reclining, (prostituted), Diana goddess image, as D Bradley pointed out a few years ago. If you're skeptical, trace all the freeways in LA with a highlighting marker, on your Road Atlas map of the city. If you know any old battle-axes who belong to Eastern Star, you may have met one or two of your local, Scottish Rite temple's 33 degree-level female ritual conductors.

Top-of-the-heap occult/corporate females are generally characterized by an exceptional desire to control others, and their glass ceiling is several storeys higher than the norm.

Hillary Clinton, Madeleine Albright, Maggie Thatcher, and Condolleeezzzza Rice are good examples. Sure, it's nice that even the color barrier is sometimes broken, but...

I hope to continue to do my part to stop all of that institutional ugliness. The 'science of the future', as Dr Reich called what we're all practicing now, relies on the input of reputable psychics, because orgone is hyper dimensional, not mundane. Our pursuit is not actually 'metaphysical,' either, because we clearly see or otherwise sense the effects, in the physical and social environment. Nothing needs to be accepted blindly any more, in contrast to what's required to belong to secret societies, which formerly represented the magical realm exclusively.

I don't think that future, non-institutionalized historians will look as favorably, on Newton and Einstein, as the What To Think Network has, until now. Did you know that Einstein blackballed Reich, among the scientific community in the early 1940s? Einstein's initial response, on seeing the dynamics of Reich's orgone accumulator, was supportive and enthusiastic and he said, 'What you've shown me will turn physics on its ear!' Not long after that, he made it clear that he was opposed to Dr Reich, though he offered no rational explanation for that sea change and refused to meet Reich or answer his letters, after that.

Did you know that Newton plagiarized the much smarter and more insightful German philosopher and mathematician, Gottfried Leibniz? He left out all of Leibniz' finer conclusions and observations, in the process, turning modern physics into an oppressive, materialistic machine, instead of a properly spiritual, technical extravaganza. True to the schizoid nature of corporate science, Newton also pursued irrational mysticism, at the time—the schizoid counterpoint to his crass materialism. The City of London later employed Darwin and Marx to further muddy the academic waters. Lord Thomas Huxley, head of Britain's espionage agencies, was their employer.

Contrast materialistic science with Leibniz' and other genuine, renaissance men's application of mathematics and science, toward uncovering and understanding hidden spiritual truths, by observing the more subtle and balanced workings of nature. Even Newton and his Masonic buddies possessed hidden, ancient knowledge about the underpinnings of the natural world, and that directly contradicted Newton's lifeless physics paradigm, but they were sworn to secrecy. Drunvalo Melchizedek lately spilled a lot of those old secret beans, in a desperate, but failed effort by his handlers to recruit young people into freemasonry. In the larger view, it's simply time for all hidden knowledge to be revealed.

Plagiarism is the time-honored tool of disinformants. The only parts of Bearden's terrifying, but oblique 'Chicken Little' presentations, for instance, that are decipherable, are what he plagiarizes from the writings of a few genuine scientists, and the only part of "THE BOOK OF MORMON" that makes any sense, is the small part in the middle of all that channeled mumbo-jumbo, which was plagiarized directly from "THE BOOK OF ISAIAH". They use the Bible a lot, instead, and why wouldn't they, under the circumstances?

In the case of Army Intelligence operative, Col. Bearden's, slick, charismatic presentations, with plenty of 'Harrumphs!' aren't a substitute for substance, and conformist Mormon 'testimonkeys' aren't legitimate claimants for that book's validity. In both cases, the subscribers may have been programmed to believe that there's something wrong with them, for not grasping what 'everyone else' in the crowd is so ecstatic about, so they enthusiastically endorse the stuff. 'Everyone else', (read: Pajama People), is governed by a deep-seated fear of being considered ignorant or faithless, by others within a group. I believe that the herd instinct is as essential for humans, now, as tails and canine teeth are.

It was a genuine, Heaven-sent miracle that James Clerk Maxwell surfaced, in the midst of London's materialistic-scientist milieu, in the middle 1800s. He prudently hid his more astonishing, liberating findings in poetry, perhaps knowing that his more explicit and enlightening pronouncements were going to be edited out of his published work before long, which is what happened. It may be that when some of the affordable, simple, free energy technology that is available is finally out on the market, some of what Maxwell tried to teach us about subtle energy dynamics will become clear.

Right now, I feel a little like someone might have felt, who had personally met Jesus Christ in His day: I have direct experience with a powerful, free energy device, and have also consistently generated and observed powerful effects in the atmosphere with simple orgonite devices, but it's a thankless task to relate that to people, who don't yet believe that free energy technology is possible, or that an uneducated little guy on the ground, like me, can directly influence the weather.

Bearden personally knew Wilhelm Muller, and had his hands on the same magnet motor/generator that I played with, and several years later, Bearden proclaimed incessantly that, 'There are no viable free energy devices on this planet.' I assume that he knows several of the other inventors, too, and I hope that by now they all realize that being schmoozed by Bearden might be the kiss of death, literally. Bearden damns orgonite with faint praise, when people ask him about what we're doing—another time-honored disinformation tactic. Since we won't seek a following, that tactic doesn't work against us; it just advertises orgonite a little bit more. I only mentioned Bearden as an example, because he's at the top of the technology disinformation dung heap, so is more widely known than the rest.

For most people who are inspired to heal the world with orgonite, it becomes obvious that we're each playing a specific part, in a larger, more pervasive and beyond-personal agenda. Anyone who wants to be a loner, in this movement, can still achieve dramatic confirmations, but they'll be deprived of experiencing synergy with others. The more we are willing to rely, unquestioningly, on our finer instincts, the more significant part we can play in this global exposition, but all of it involves 'gifting.'

D Bradley, a phenomenally talented and highly trained psychic, who popularized the word, 'gifting,' for this network's main activity—distributing orgonite in the environment—also once wrote in a forum post, 'The Operators are standing by, to take your call,' and 'The Operators,' too, quickly turned into a banner phrase.

This is the fellow who almost single-handedly banished the smog and poisonous chemtrails from most of the Los Angeles Basin, and brought abundant rainfall to the region, by transforming countless thousands of new, deadly transmitters there. He also located and gifted key institutional occultists' properties, and satanic energy grid benchmarks and distributed around two score of orgonite cloudbusters in LA and beyond.

DB produced the documentary, "CHEMTRAILS: CLOUDS OF DEATH", which is available for no charge, as a download from www.ethericwarriors.com. It's a no-holds-barred description of the biological weaponry found in congealed lumps of the sky spew, which more quickly fell to ground and was analyzed—at tremendous personal risk—by independent, small labs in the US and Canada in the late 1990's.

The film encourages the watcher to make or buy an orgonite cloudbuster, the demonstrated solution to the chemtrail dilemma. DB is, hands down, the most fascinating, enigmatic person I've ever met and I've met a lot of interesting people in 57 years of inquisitive living and traveling, here and abroad. The enemies of humanity are doing their best to erase him from public awareness again, by the way. His current website is www.thevine.net/cbswork.

'Orgone' is Dr Reich's term for 'life force,' which is the energy matrix of the physical and, (apparently), non-physical universe. Matter and energy are expressions of orgone. Of course, there's nothing 'solid' in creation. Dr. Reich, too, was nearly erased from public awareness, by the time of his untimely death in a federal prison in 1957, a day before he was to have been released.

Carol and I invented the orgonite cloudbuster, in the winter of 2000 to 2001, based on combining the earlier, very successful, cloudbuster designed by Dr Wilhelm Reich in the late 1940s, with orgonite to make it safer to use for both the operator and the environment. Reich, himself, explicitly advised people to only use his cloudbuster design intelligently and sparingly, in order not to create severe energy imbalances in the atmosphere or injure the operator, but an orgonite cloudbuster can be safely left standing in the environment, without risking harm to anyone. Also, it can be safely handled, any time.

In July 1998, one of my zapper customers had told me that Karl Welz was making and selling devices made of epoxy and metal particles, and that it apparently generates orgone energy. Welz later began calling this simple material, 'orgonite,' after Carol and I began widely sharing the material's ingredients and some of its applications on the internet, in 2000.

I had been making orgone accumulators in various shapes

and sizes, since the middle of 1997. Various psychics and energy sensitives, including Carol and, previously, my own nine-year-old daughter, Nora, observed the energy dynamics for me in specific applications. I was fortunate to find and read Serge Kahili King's small, but substantive book, "EARTH ENERGIES", in the summer of 1997, because King advises making orgone accumulators by layering aluminum foil and plastic food wrap. Dr Reich's recommended materials were considerably bulkier, heavier and more costly—perhaps because he had to work with material that was easily available in the 1930s. King's recommendations enabled me to create a wide variety of orgone accumulators for various applications, including one that could be worn under the shirt on the torso to aid dowsers, several hollow pyramids with removable doors, some cubes, a tetrahedron and even a very big cabinet, which I could lay down in.

For my second psychic reading from Carol that summer, I traded a six-inch, cube orgone accumulator, which she said suddenly and dramatically enhanced her ability to see energy and receive psychic impressions. I paid for her first reading, a month before, with a zapper, which she was still using when we finally got together, three years later.

Orgone accumulators, which Dr Reich invented in the late 1930s, don't differentiate the quality of the energy they absorb, which is why he built his lab in a remote location. We found out that orgonite not only absorbs and condenses healthy orgone, but also transmutes destructive energy into positive life force in the process. This is why we casually call orgonite devices, 'energy generators.' That's a small mistruth, but it's harder to wrap one's brain around 'Orgone Regenerators.'

When Carol and I finally get a place, (which mainly means, to me, that I'll finally have a proper shop to work in again), I'm going to make another grand-scale orgone accumulator, built for two, because a few minutes in one of those is quite rejuvenating—I've never achieved that energizing effect with orgonite, in fact. A small amount of simple orgonite on or beside any accumulator, guarantees that only positive orgone radiation will be accumulated.

Nothing is being generated by orgonite; bad energy is simply being transmuted into good energy, and adding some crystal to the mix amplifies and focuses that effect. Crystals have an innate tendency to organize ambient energy, and orgonite, without a crystal added, produces life force in a slightly chaotic, but still-friendly form. Powdered quartz, rather than distinct crystals, causes the energy from orgonite to be more erratic.

One of the effects of the transmutation process, is that it influences human attitudes and behavior within its energy field, in a positive way. A typical, 12-ounce orgonite device extends its influence over an entire household, and a three-ounce piece of orgonite per square block, throughout a large neighborhood, creates a dynamic, synergistic energy grid, over the entire area and beyond. 'Gridding' is one of the ways we can achieve more, with less orgonite. Orgone fields within homes are generally much more intense, than orgone fields in the atmosphere. When Georg and Trevor gridded Soweto, outside of Johannesburg, they changed that murderous, hellhole into a pleasant community, in a short time. When they returned, a month later, the trash had all been removed, a lot of trees had been planted and people's countenances and behavior were pleasant, even toward the white visitors. When they were tossing the orgonite, they didn't dare stop their car.

The synergistic effect of grid-gifting has been well demonstrated in metropolitan Toronto, too, where 25,000 or so of these simple devices were distributed pretty evenly, among twenty million inhabitants, by Steve Baron and crew, in the past couple of years. If you spent time in Toronto, before two years ago and will return there, you're going to witness an incredible transformation in the ambience and the once-smoggy atmosphere.

Lake Ontario's water, along Toronto's shore, is pristine and very clear now, but Steve did that with only a few pieces of orgonite. Three years ago, he casually tossed them out from the shore, here and there, before he and his friends initiated their more extensive urban and suburban gifting agenda. Local environmentalists, who had campaigned to get the water cleaned up, haven't figured out how that suddenly happened, though this has been discussed at length on the internet for anyone to find, with a little googling. I think the environmentalists assumed they had to induce the government to hammer nasty, polluting corporations in order to help the lake. Most environmentalists are just PJ folks, who dream that they're awake in a storm-in-a-teacup world, which they have created. In their dream world, governments and corporations are magically separate, and lawmakers are good people.

Institutionalized environmentalists are generally opposed to what we do, because they object, out of hand, to our use of fiberglass resin. The ship, 'Rainbow Warrior,' probably has very toxic bottom paint, and Greenpeace, no doubt, uses mostly fiberglass boats, too, but intellectual integrity has never been a hallmark of Petroleum-Cabal-funded environmentalist organizations. Catalyzed, cured fiberglass resin is inert, of course—no more harmful to the environment than beeswax is.

For many years before, people in Toronto were explicitly warned not to swim in those waters, which had become quite murky as well as poisonous. In fact, Steve had to abandon his lucrative parasailing business ten years ago, because the water had became too polluted for water-skiing.

In Idaho we lived near Lewiston, in the Snake River Valley. We dreaded going to that city, thirty miles south, to buy our fiberglass resin, (the alternative was Spokane, seventy miles north), because as soon as we descended into the deep gorge from the Palouse, (high prairie), where we were living, we entered a palpable wall of stinky, eye-watering, brownish yellow smog from Potlatch Mill. That pulp mill was a rallying point for the environmentalists in the region, who had tried for many years to get the plant to stop spewing that turgid, opaque mass into the atmosphere, and literally filling up the entire gorge with it for fifty miles, downstream. The environmentalists were entirely unsuccessful, of course, and it might be because the corporation, who owns the mill, also owns much of the state government, and they're certainly in league with the corporations, which fund the environmentalist organizations. Interlocking directorates was a nifty way for the City of London to make one, global mega corporation look like many little competing ones.

It's a good thing that so many people want to heal the earth, but it's obviously not good when one assumes that it's necessary to support a questionable organization, increase the size of centralized governments or subscribe to an artificial, new bizarre ideology, to get it done. The popular ideology among environmentalists, these days, is based on a hatred of humanity, who are openly called parasites—a virulent infestation on 'the goddess,' Gaia's, delicate hide. If intellectual integrity prevailed in these organizations, you'd expect them all to suicide, as a demonstration of their devotion to Gaia, but apparently they don't consider themselves to be intolerable affronts to 'the goddess', like all the rest of us apparently are. I think one can expect the pendulum to swing the other way a bit, after six thousand years of bloody misogyny, but I hope we can get past this silly, goddess stuff, as soon as possible, and just start practicing gender equality in a productive, balanced way.

An old witch friend of Carol's in Lewiston, Laurie, agreed to host an orgonite cloudbuster in her yard, (about three miles downwind from the mill), and from the day we set that up in March 2002, the atmosphere in Lewiston has been crystal clear, nor does one even smell the mill any more, unless one is very close to its settling ponds. The environmentalists haven't mentioned the improvement. Maybe, in their dream world, the gorge is still plagued by stinky smog.

Most of us around the world, in this network, 'gift on the sly,' careful to keep out of sight of the ubiquitous FBI, CIA, NSA, Mossad, MI5, MI6, KGB, Church Universal and Triumphant, Scientology, Ramtha, Mormon and other occult/corporate and quasi-governmental spy agencies, who send trackers on their corporate masters' behalf to try to find out where we're putting our gifts, so that they can remove them, later on. The less discerning, lonesome gifters are so happy, when clever saboteurs from these agencies befriend them, that they invite them along on gifting sorties, though that's happening less and less. In that case, the orgonite is routinely retrieved and destroyed. Now that there are no remaining open membership, quasi-viable internet forums in the movement, the saboteurs are finding it much harder to infiltrate the network, and people are generally savvy enough to know not to drop orgonite, when anyone is watching them. Keeping in motion and being aware of our surroundings, pretty much guarantees that pavement artists from the various agencies won't find and remove our gifts. One well-meaning fellow, who was an active Monarch asset, gifted an entire large city, but psychics were looking out of his eyes throughout the campaign, and the CIA's sewer rats routinely removed them all. He got clear of drugs, hence got clear of the Monarch Program, and did the city again and this time it apparently took.

Ultimately, the best way to ensure that orgonite will get thoroughly distributed, is to hand it out to people for their own homes, but in the developed countries—and I assume it's creepiest right now in Russia and China—most people are just too paranoid to accept anything from strangers, these days. Europoid Americans are mostly just terrified that you'll try to sell them something, and they only respond to predatory salesmen; not likely to sincere gifters, like you and I. You can gift their neighborhood on the sly, then go back a week later, and I bet they'll be happy to accept some orgonite after that, because the edge will have been taken off their paranoia, by the orgonite in the neighborhood.

The African gifters mainly hand out orgonite devices to people, rather than hide it in the environment, because most of the people over there are not conditioned to distrust every stranger, and they seem to instinctively understand what orgonite can do for them. If you can afford it and have the time, go to Africa and hang out with some gifters there, (take along crystals and buy resin for them), and you'll directly experience that liberating phenomenon, as some of us in the West have by now. Some of them post on www.ethericwarriors.com, and you can contact them from there.

Chapter Four

Treating Global Treason at Home

A more personal benefit of gifting, is that it helps us feel less resentment toward the retreating, parasitic world order, because we can immediately see the healing effects in the environment, from our tower-busting efforts, and we know that we're destroying their essential death-energy matrix, at the same time. That gives us a sense of satisfaction, as you can imagine. The economy of waging war on the satanic world order this way, compares to taking down a billion-dollar B-2 bomber with an arrow.

I focus on the subject of conspiracies, for now, because the occult/corporate hierarchies are mainly in the business of ensuring that you stay asleep and unaware of your destiny and potential. They've been phenomenally successful until now, thanks to monumental conspiracies that span millennia, in some cases.

Sharing a glimpse of 'Them' with you, if you're one of the unwitting sleepyheads, (as opposed to Pajama People who clutch at sleep), can induce you to catch a glimpse of the power that's latent in yourself. That can be pretty inspiring, and it can lead you directly to constructive action. It's okay, if you initially feel paranoid as hell. We all go through that, on the way to self-empowerment, of course. When I got my first glimpse of the way things are, I was in Tonga, in 1984. I had just gotten a copy of Stan Deyo's, "THE COSMIC CONSPIRACY", and the wife and I stayed up all night, reading it aloud; spellbound.

I'd just gotten a publishing job in Fiji, and had finished a sign-writing project in Western Samoa, so we waited five months in Tonga, (with our two small children), for a work visa from Fiji, which never came. It was the only time in our 22 years together that we had a semblance of a marriage, sad to tell.

By the time we finished that book, which is heavily documented, we were convinced that someone was watching us. We both laughed that off, but in the past five years I've become acutely aware of how thoroughly the intelligence agencies, and their cult subsidiaries, use psychics to hunt in the ethers for people who are in the process of waking up. The psychics in our own network tell us that the awakening ones show up as blips, on a cosmic radar screen. There are protocols in place, to 'visit' these awakening ones and induce paranoia, if their What To Think Network programming is relatively intact, and to take 'further measures', if they persist at unraveling the What To Think Network's dreamscape. As I mentioned, this parasitic, psychological infrastructure isn't solidly in place, in the developing nations, which is why you might feel an uncharacteristic sense of personal freedom, when you go to those countries, if you were raised in the West.

They're amazingly skilled and thorough. Maybe they were counting on AIDS, endless, (corporate sponsored), violent strife and their other new biological weaponry, to prevent the empowerment of humanity in Africa.

That book is two-thirds full of substantive, documented material, and one-third full of disinformation, which is conveniently separate from the good stuff, so can be ignored—typical quasi-Christian doom-saying, with the author's thinly-disguised claim to be a prophet, and inferences that if you play your cards right, you'll be with him and Jesus, when the $#!+storm begins. The book was written in the seventies, but the documentation about anti-gravity technology and conspiracies is solid.

Disinformation in a book, doesn't seem to reach out and fog the mind and darken the heart, the way disinformation on the internet and television seems to do, maybe because you can skip that part in a book. There's probably a lot of subliminal programming coming through the monitor, when you're on a disinformation site, like there is, when you watch The What To Think Network. Nobody's immune from that energy-draining technology, so it's better to just avoid it.

When we got back to the US in the middle of 1985, I was exposed to more objective conspiracy information, including, "WORLD ORDER; THE HEGEMONY OF PARASITISM", by Eustace Mullins, and "NONE DARE CALL IT CONSPIRACY", by Gary Allen. I found most of the material, by listening to a fundamentalist Christian radio station, while at work. I was still painting signs in those days. Most of the guests on those talk shows were disinformants, but I was able to profile them, so when many of them showed up on the Art Bell Show, in the next decade, I already had them pegged.

However deep you suspect the occult/corporate conspiracies go, assume that they go a lot deeper, and you'll have a balanced, rational view of the dynamics of occult/corporate hierarchies. Disinformants, especially on the internet, routinely expose one level of a conspiracy, in order to direct the subscriber's attention away from a deeper level, in fact—yet another tired and tiresome Machiavellian ploy. Loyalty and team spirit don't mean the same thing to them, as it does to you and me. If you want to understand them, you would need to be a parasite or predator, which is why I don't seek to understand their motives. I guarantee that nobody becomes a predator, simply because his mommy didn't buy him any toys. It's always a conscious choice.

Identifying the other side's playbook is empowering, but only if you've got some empowering solutions, and are capable of adopting a strategy to defeat them. What we're all doing in this global network to end tyranny and heal the world really works, which is why we're all doing it. It's also why we don't angst over Chicken Little doom scenarios. I want you to feel the same job satisfaction and sense of accomplishment, that we've been achieving. Anyone can do what we're doing, and if that weren't so, there wouldn't be a global, grassroots network; we'd just be a few chest-pounding, but ineffective dilettantes.

For instance, with a cheap, three- or four-ounce, homemade, orgonite towerbuster, you can disable, with impunity, your neighborhood's million-dollar, militarily fortified, deadly-orgone-radiation transmitter. You can easily make the towerbusters in muffin trays in your garage, fifty at a time, from a gallon of cheap fiberglass resin, a small pail of waste metal from a machinist, some muffin pans from MalWart, and a handful of cheap crystals. Then you can disable all of the new death transmitters, in and around a town of 50,000 or so people, with what you just made.

Disabling a death tower with a towerbuster, is like shooting a bomber down with an arrow, or disabling a tank with a slingshot, and the odds are in our favor now, which is another way of saying, 'The meek are inheriting the earth.'

> Meek: quiet, gentle and submissive.

I'm quiet and gentle - ask anyone who knows me - and I'm happily submissive to the agenda of The Operators. I'm not a Muslim, but 'Muslim' in Arabic means, 'one who submits' and this is in the context of submitting to "the God's", (Allah's), will so I think it's good to mention that. I look for opportunities to present Islam in a fair light, as another effort to counteract The What To Think Network's ongoing, vehement xenophobia campaign.

Most of the people who do the lion's share of the gifting and predator-blasting work, are like me that way, though of course we're a quite diverse network. The day of the alpha personality's dominance in human affairs is now past—gone the way of tails, canine teeth, and televangelists. Ayatollah-you so!

Gifting's simple but powerful dynamic, actually goes a step further than disabling the transmitters; the orgonite perpetually changes the transmitter's deadly radiation into life force, beyond a very slim zone around the tower itself. Psychics consistently see this dark energy zone extending only a few feet beyond the tower, in fact, even though the orgonite towerbuster that's making it happen is up to a quarter mile away. Maybe someone will figure out why that happens so consistently, and can then explain it to the rest of us, in simple terms.

Genuine communication technology doesn't produce dark energy fields, and orgonite never interferes with legitimate communication technology. All standing towers, radomes, dish arrays and building top transmitters are now either used to harm the population, or harm the atmosphere. Radio and TV stations' transmitter towers and dish farms are now also used in the HAARP network, for instance. 'HAARP' is weather warfare against the population, usually waged by one of their own government agencies. The claim that this is all done from a few remote locations, including Anchorage, Alaska, is a Chicken Little misdirect, designed to stop you from noticing the less obtrusive, but more offensive, omnipresent, HAARP tower arrays in your own community, and along all of the primary highways and coastlines. The death towers aim deadly orgone radiation directly at people. They interfere with cell phone reception, which shows that they aren't set up for cell phones.

After gifting your community, the net result of the erection of the death-transmitter and HAARP networks, rather than serving the genocide agenda as 'They' intended, is that the overall ambience of the atmosphere is enhanced to a level that's superior to what it would have been, if the towers had never been built. The death technology becomes life-force generating technology, after you gift it all.

The scientists and technicians, who are employed by the retreating world order, are unwittingly sabotaging the destructive agenda of their employers, by erecting these towers, rather than enabling it, in other words. They dare not simply throw up more and more towers, after we disable the ones that are in place, because there are more than enough of these, right now, to eventually shock, even the most committed sleep addict, into awareness of the obvious nature of the towers.

The 'new world order', announced by Reichsfuhrer Bush Senior, in 1990, is simply the licentious, titillating, momentary display of an ancient parasitic order, of course, and you can probably see that they jumped the gun by announcing themselves, back then. Note that the UN was still being presented as the model for globalism, then. The UN has rather become like an embarrassing, loquacious, alcoholic uncle, in more recent years, or a constantly masturbating pet monkey, at a dinner party, and it's rarely mentioned by the What To Think Network, any more.

'The New World Order!' potentially exposed their true agenda to the Pajama People eleven years later, by suddenly erecting millions and millions of brand-new, death-transmitting towers, around the world, in the months following their demolition of the World Trade Center. The HAARP network had been in place for several years before that, of course, but was suddenly expanded after the destruction of the WTC.

Have you noticed that nobody is prepared to aggressively defend the notion that these towers are for cell phones? Since picture phones are blatantly advertised to work from the satellites, how much more of the cell traffic do you reckon is also handled by cheap communication satellites by now, rather than by million-dollar, fortified, secret-tech, neighborhood transmitters? The French, (probably several other nations, too), have been shooting inexpensive, communication satellites into orbit, with cannons, for decades, while the What To Think Network, has been directing your attention to the Great Boondoggle of the Skies, NASA's decrepit space shuttles. NASA has developed a reputation, since the Apollo scam, of being as innovative and pioneering as the US Post Office, so nobody takes them seriously any more. That's not a dig at postal workers; Carol and I operate a mail order business, and we prefer the US Postal Service to the commercial carriers, because they're more reliable and conscientious. I guess it's too bad that they don't also run NASA, now that I think of it.

The technology of the towers, is a better kept secret than the Manhattan Project was, and the massive bundle of cables that run up each mast to the transmitter rods, dishes, horns and panels, stand in outrageous contrast to the slim, little power line that may, or may not, come from a nearby pole to the fortified concrete bunker, next to the overbuilt tower.

Following the rabbit hole further down, we discover that there are unregistered, nuclear power plants, all over the landscape. The new cooling ponds mark the locations of these little, underground reactors. If you live up north, it's especially evident that the ubiquitous, rectangular double ponds, gravel lined and surrounded by a chain link fence, are for cooling, because they're yellow-green in color, from antifreeze. I hope you'll keep your eyes peeled, because you're going to find them, if so. I also hope that you're not so distracted by the lack of any reference to these, by the What To Think Network or even on the internet, (except for the consistent reports on Etheric Warriors, by people in the gifting network), that you won't go looking for the ponds.

Go find them and gift them, okay? Do your part and gather the requisite, visual confirmations in the atmosphere, after you shut down those little nuke reactors! Residual countryside smog will quickly disappear in that area, and you're likely to see some sylphs in a blue sky. The sylphs seem to be quite eager to show us their approval and encouragement, these days.

When you find an unregistered, nuke cooling facility, take note that each of the twin ponds alternately empties and fills. While you're there, toss a towerbuster into either pond, and when you return, you'll see that both ponds are full, which means that the reactor underground has been de-activated by the 'orgonized', cooling water. If you're psychic, note that the reddish brown, deadly orgone field, that existed before you gifted the pond, is mostly gone, when the ponds are both full. Reddish brown energy is characteristic of the deadly energy field, generated by an operating nuclear reactor.

The underground reactors provide most of the power for the new tower network. If the new towers were powered by the grid, your electric bill would have suddenly doubled, right before 2002.

Nobody knows for sure whether turning on the towers full blast, which may be what is required to throw the surrounding population to the ground with seizures, would overcome the effects of the orgonite. I suspect the other side has already determined that this is not so, otherwise maybe, we'd have heard of a case where everyone, within a mile or so of a single tower, suddenly fell to the ground, incapacitated. They always experiment on the populace, of course, and that usually finds its way into the news, onto call-in talk shows, or the internet. In the early nineties, for instance, I heard many reports of chemtrails in isolated areas, and of people getting deathly ill, after handling congealed goo that fell from the sky. Californians are favorites for testing biological and chemical weaponry, over a wide region, because—let's face it—they're more self-indulged and vacuous than most of the rest of the country are.

Weapons testing on Californians was even admitted by the government, on the What To Think Network, also weapons testing on rural Black communities in the South. They even showed the Congressional legislation that made it 'legal,' and I think they were just passing the buck, in that case. Everybody assumes Congressmen are predators, after all, and the Federal Reserve Corporation's court system made it impossible to sue the government, after the end of the rule of law in America, in the mid-1930s.

As with so much of the rest of the world order's genocide agendas, we had to operate on the psychics' intelligence-gathering efforts, in order to develop an effective strategy to disable these new transmitters, in a timely way. I say, 'timely,' because the reputable psychics were all seeing that the ultimate purpose of the non-HAARP towers was to throw the population to the ground, on command, so that relatively few foreign troops could round up the dissidents, (you and I), for fast execution in the local and regional facilities, which have existed, fully staffed, since just before the Waco massacre. There are at least two per state, strategically located, and they're also throughout Canada and, presumably, the rest of the world. There are many more holding facilities—at least one in every small town, commercial district and large neighborhood. These are identified, usually, by the way the barbed wire panels at the top of tall chain link fences are leaning—inward in this case, to keep people in. Fences that are meant to keep people out have that top panel leaning outward, instead. Really bad places, like the combined facilities of Earthlink and the West Coast NSA Headquarters, have fortified gates, coils of razor tape on top of tall fences, and guard towers on the corners.

D Bradley was taking a manuscript to the publisher—an illustrated expose about the North American death camp network—one day in 1999, when he was shot through the throat by a sniper, from across the street, and drowned in his own blood. Thankfully, his family got him to a hospital in time to revive him.

The reputable psychics told us that the chemtrail program's agenda was to sufficiently weaken most of the population, with non-lethal bio-weaponry. After enough of the population was chronically ill, they would then apply the aerial kill-shot: anthrax, ebola, smallpox, etc. This was to have been in process during the time the towers were erected, and apparently Muslims would have been blamed for it all. Fortunately, so many cloudbusters were being built here and abroad, during the previous six months, that huge holes were literally being made in their designs. Before the following summer, most of North America and Western Europe were within range of orgonite cloudbusters, which quickly neutralize bio-weaponry, long before it reaches the ground. Perhaps this is partly done by simple ionization, but certainly it's due to massive infusions of positive orgone radiation, into the upper atmosphere. We tracked the range of an early Canadian cloudbuster's ability to disappear chemtrails, at around 250 miles, but I think the range varies a lot according to several conditions.

If the chemtrail, AIDS and death-tower agendas had worked, I'm sure the population of the world would have been reduced to half a billion by now. The bottleneck of the parasitic world order is apparently personnel; there are only so many people in the world who are willing to serve a destructive agenda consciously, after all, and I suspect that there are an equal, or greater number of people who are motivated in the opposite direction. Judging by the rapid expansion of the informal gifting movement, I'm guessing that the number of good guys may greatly exceed the number of villains, in fact. Most Pajama People will do almost anything to collect a paycheck, as long as it's not blatantly predatory and disturbing. Look at all the 'nice people' who are cops, or work in hospitals, for instance. The twin, massive American gulag archipelagos—the penal system and hospital networks—are stark testimonies to the deeply schizophrenic nature of our society, at the moment. Prisons are filled, mainly with non-criminals, and hospitals are filled with sick and dying people, who could mostly be quickly cured with a circuit that costs a dollar to make.

The vast majority will go along—within limits—with whatever agenda they're subjected to. Destroying or even distracting the genuine activists and people of conscience would tip the scales in favor of the parasites, but that many PJ folks, most of whom are always potential activists and people of conscience, would still be troublesome, of course, as soon as they woke up, so their numbers, too, would need to be drastically reduced by rapid attrition, hence the biological attacks, which could conveniently be blamed on terrorists, just like the demolition of the WTC and the Gulf War records section of the Pentagon—'the kickoff' for the Homeland Security Abomination and a raft of treasonous, draconian legislation—was, on September 11, 2001.

The new, deadly technology seems to require the establishment of a matrix of deadly orgone radiation, and malefic, operative will, in order to function. Orgonite constantly absorbs, then transmutes deadly energy, (including expressed malice), into positive energy; life force. It takes more orgonite to fix deadly energy sources, among the populace than it does in the wilderness, the skies or in the oceans. An individual with malicious intent generates an enormous volume of deadly orgone radiation, even unconsciously sometimes. A place like Los Angeles, where such a large percentage of the population are predators, is a pretty tough target for gifters. It's a lot easier to clear up the atmosphere above the clouds, which is where chemtrails are spewed.

The insertion of parallel pipes into a two-gallon mass of orgonite, draws disruptive energy in, from a great distance in the upper atmosphere, and instantly sends the restored version right back to where the bad stuff came from. Psychics have seen this process as a sort of reverse, rotating tornado, generating an enormous canopy that extends up to two hundred fifty miles, depending on conditions. The field is much smaller in a city, much bigger in the wilderness, for instance.

For healing the lower atmosphere, we need the smaller devices, more widely distributed.

In those early days, the only effective psychics who were available all the time were Carol and D Bradley. Other good ones, and a fair share of charismatic but crooked ones and their dupes, came and went, but these two stuck with the network in those potentially dark, early days. Now there are many more good, reliable psychics, including Laura Weise and her husband, Dr Steve Smith, in Montana, Tracey Ann Doran in UK, vo Joanna in Brazil, Kelly McKennon in Washington State, Dr Charles Olson in the Phillipines, Kizira Ibrahim in Uganda, Denis Couture in Quebec, Lynda Frey in Georgia, Dr Dirk Verelst in Amsterdam and some German, Dutch, French, Italian and other psychics in Europe, who mainly participate on the non-English language orgonite boards.

Carol and I developed and thoroughly tested some simple

protocols for disabling the new towers with orgonite, during the summer of 2002, which we soon shared with thousands of people, and many of those thousands are sharing them with thousands more, genuine activists around the world, including many sober youth, finally. For a long time, anyone who found this material on the web was awestruck to meet others who felt as they did, because, let's face it, genuine activists are quite rare in the West and are usually ostracized by their own families, and nearly everyone else they meet, (the PJ majority).

Keeping it simple guarantees that anyone can do the work, achieve success and acquire visual/sensory confirmations immediately afterward, and from then on.

These terrorizing eyesores are even more 'in your face', than the chemtrails were. In fact, most people, who now see chemtrails in the present, innocuous form, never noticed them, when they were quickly whiting out the sky, destroying the cloud cover, killing thousands of infants and old folks, and sometimes leaving a pathogen-laden, kerosene haze on our windshields. Lots of newcomers were so unaware of the new forest of death towers around them, for the past four years, that they assume that the towers were suddenly erected last week. Movies that show footage of the 'HOLLYWOOD' sign, carefully edit out the massive tower array that's just uphill from the sign. You can see it, and if you visit LA, you'll be shocked to see how many death towers were thrown up on that hill, since your last visit, if that visit was before four years ago. You'll be astonished and refreshed by the improved ambience and lack of smog, thanks mainly to D Bradley's tireless, gifting campaigns in the LA Basin.

The little smidgeons of chemtrails that remain in the sky these days, now that the massive, international genocide agenda has been disabled, do little more than seed small clouds and you'll likely see sylphs show up and literally gobble the chemtrails. Meanwhile, Bearden and a few other disinformants, are telling anyone who will listen, that things are actually getting worse and worse. Pravda's Sorcha Faal has gone a step beyond Bearden, and is telling people that the world, (especially the US), is now being rapidly destroyed by massive cyclones, earthquakes, famines, meteors, civil wars volcanic eruptions and even global-scale warfare. She actually has a following in the US, who apparently believe her, in spite of abundant evidence that Americans are experiencing a reducing crime rate, lovelier skies, more moderate weather, better health, more prosperity, less drought, pleasant urban ambience, almost no smog, less general anxiety, no 'terrorism,' and generally a better life. I don't think her followers in the US get out very much. Maybe they've all become troglodytes, whose only connection with our world is via their modems.

As often as not, sylphs can be seen in the sky, 'eating' the remnants of chemtrails and HAARP muck, in fact. The reappearance of the sylphs, after many generations of absence, may be the most positive affirmation of this unassuming gifting network's success. Seeing and recognizing a sylph's distinctive, magnificent, cloud sculpture, initiates an immediate heart bond between the viewer and those visually ephemeral, but curiously enduring, benevolent creatures in our sky. Nobody seems to know who or what they are. My wife calls them 'angels' and others say that they're big, benevolent elementals. Maybe we'll find out pretty soon. They were not uncommon, before the early 1900s, and then were rarely seen after that, until a couple of years ago.

One of our loyal, zapper customers who owns the Ballanchine Ballet Companies in California, told me that when she first saw one of the Sylphs in the sky over Berkeley, a couple of years ago, she was particularly heart-struck, because when she was young she had danced in the ballet, Les Sylphides.

The activity of the newly returned Sylphs in our skies is a slow ballet, so we need to pay attention for a while, to appreciate the dance. Someone will no doubt film this slow-motion process for us. Www.ryanmcginty.com has some good sylph photos, and several are also posted by individual gifters on www.ethericwarriors.com.

Disinformation websites like, www.rense.com, began claiming that the chemtrails are going to kill everyone. This started a few months after it became clear that the chemtrail program was essentially neutralized, and it was several months, before he began warning people not to make orgonite cloudbusters. Before, when the chemtrail program was an imminent threat, and was killing thousands of old people and babies, little mention was ever made of chemtrails, by the disinformants. The disinformation agenda about chemtrails is slowly shifting from doom-saying to claims that the chemtrails are 'stopping global warming.' Since most folks who visit disinformation sites aren't critical thinkers, they don't reflect on the fact that chemtrails are mostly spewed over population centers, on a few continents.

When Dr James DeMeo traveled to Germany for a lecture tour in January 2004, he explicitly warned his audiences to stay away from orgonite, especially 'Croft-style cloudbusters.' Before, there were only a handful of orgonite cloudbusters in Germany, but right after he left, a great number of them were built and deployed in Germany, Austria and Switzerland. The sudden celebrity of orgonite there formed the basis of the German language orgonite forum, which is hosted by Georg Ritschl, a Berliner émigré, who hosts www.orgonise-africa.net Many stores in Switzerland are selling orgonite devices now.

It may be that the Pajama People are not yet fully aware that their emperor has no clothes, or they're programmed to only look at his blow-dried coiffure, rather than at his fat, pimply buttocks, but we at least gave that stinky emperor a terrific wedgie, in the latter half of 2002, when reports started circulating on the internet about how easy it is to disable the death towers, with simple orgonite.

The next step, after the PJ folks wake up a little more, or lose interest in paying unlawful taxes, or otherwise supporting parasitic, corporate tyranny, may be to dissolve the occult/ corporate world order and hold them all accountable. I think that properly lawful and representative, minimal government, throughout the planet will follow as naturally as whiskers suddenly start to grow on young men's faces and their voices deepen.

I'm tossing out a lot of suggestions, but I'm confident that tyranny will end, one way or another, and the eventual solution will be more creative and empowering, than any of us can imagine at the moment. I just want you to reflect on some of the possibilities, as I'm doing, and talk about it openly with others, because that's how every great undertaking gets started. All good things in this world come from spontaneous, timely, grassroots movements—you and me—not from dried-up, corporate, state, academic or religious hierarchies.

One function of human law, as an expression of universal law, is to sanction the destructive behavior of others, so that everyone else can have open and abundant opportunities to do and become whatever they wish. The little governments we create, need to reward good behavior as well as punish predatory behavior, and since most power needs to reside in local governments, we can have a voice in that process. Watch how fast people will leave the big cities, so that they can participate in government. The internet and good transportation networks make it unnecessary for us to congregate, by the millions, in one place.

As long as nobody is being exploited or harmed in the process of self-empowerment, how could that be considered unlawful? We middle class Americans have to be considered lawbreakers by the treasonous courts, in order for us to earn a livelihood, but we know that's what our own economic freedom requires, otherwise we'd let this treasonous corporation tax us into poverty, as is happening in all of the other National-Socialist countries now.

Twenty million people in America each year don't file income tax reports. That's a pretty big chunk of the middle class, I'm happy to say, and I bet most of us are self-employed. As long as the black market is the only free market, that's where we'll continue to conduct our own business, along with twenty million of our compatriots. When corporate tyranny is destroyed, so will the black market instantly disappear, and millions of men and women who are unjustly imprisoned for non-crimes, often tax-related, will be set free. Irwin Schiff was recently thrown in prison because he proved, beyond a doubt, that federal income tax is entirely voluntary. They couldn't defeat him in their courts, so they threw him in prison to shut him up. Let's see if 20 million became 30 million, on account of that. Extortion Day, April 15, just passed so the new figures might come out soon, if the numbers aren't considered a direct terrorist threat, by the Homeland Security Abomination.

What we do in the informal, intel-gathering, predator-blasting sessions should be considered the norm for policing tyranny, throughout the world. After all, it's more expeditious and less messy, than shooting the government predators would be, and it's done entirely within the law. Why rely on cops for protection, when we can do it so well for ourselves, mostly? Cops, perhaps especially these new jackbooted Gestapo guys with shaved heads, battle gear and attitude, only show up after a crime has been committed.

The courageous cops in New Orleans, who tried to protect the Black populace from murderous Homeland Security Abominations, after our federal government blew up the levies and flooded the city, were genuine heroes, killed in the line of duty, and that can't be mentioned enough.

We 'show up' before most the larger crimes are repeated, thanks to the psychics' ability to gather intel, and we deal with the perpetrators before they have a chance to achieve their destructive ends. We missed a beat with the New Orleans massacre, unfortunately, but Robert Schoen's cloudbuster at Lake Ponchartrain, at least diverted that HAARPicane from New Orleans, in time to prevent strong winds from muffling the explosions at the levies. Everyone in town heard that and knew it was sabotage. Apparently, they picked New Orleans to try out some of their martial law tactics, because it's an isolated city. It essentially has a vast moat around it, in the form of impenetrable swamp, and only a few roads, easily blocked by the Feds, go into the city. Even MalWart tried to help the survivors, but their truck convoys of aid were turned away, by the Homeland Security Abominations, who probably would have looted them, if people weren't watching.

I shudder to think what the Abominations could have committed in New Orleans, if the HAARPicane hadn't been diverted by Robert's cloudbuster, in time.

Carol and I are doing our part in disabling the Southern Coast regional HAARP network, to ensure that this won't be repeated, ever again.

Before long, I don't think that the institutional exploitation of others will be possible any more, thanks to this new, grassroots, etheric, policing practice, which anyone can take part in, and which reputable psychics enable. The US Constitution sanctions 'citizen's arrest', and this is essentially what we've all been doing to the parasitic hierarchy, for the past four years.

Criminals will always need to be also physically restrained and punished, though, so we still need lawfully established police, courts, and prisons, at least for a while. The end of the rule of misogynists will enable mothers to train their little children to have good characters, because motherhood will be seen as a genuinely honorable occupation again. When the corporate hierarchies are forced to release us all from their economic chokehold, a family will once again be able to live comfortably on one salary. I don't think we'll need to force them to let go; I rather assume that they'll lose their grip, due to the dissemination of free energy devices, grassroots and enlightened agriculture, zappers increasing on the internet and other decentralizing technology. Orgonite in their presence will continue to cause these predators and mega-parasites to continue to lose their focus, and that can't be underestimated.

Mao's anarchic Great Leap Forward, created a generation of motherless, illiterate criminals in the '50s and '60s. The popular, grand-scale western democratic liberal practice, of turning infants over to underpaid, ignorant and vindictive day care center employees, threatened to do the same thing in the West, by feeding the massive, CIA/MI6 Monarch and Tavistock Program, countless millions of little, involuntary recruits. We might yet squeeze through this present social calamity, relatively unscathed. I said, 'calamity.' How much more destructive is this benighted family paradigm, than a mere natural disaster or war could be?

Many among the daycare generations, which vastly swelled the ranks of the satanic, dissociative Monarch and Tavistock agendas and the US prison gulag, are waking up, against all the odds, and that's been tearing the Monarch and

Tavistock agendas apart, thank God. The bad guys always underestimate the power of the human spirit to lift itself up, out of bondage and hopelessness. They underestimate it because they can't understand it.

According to the Uniform Commercial Code, which silently supplanted the rule of law in America, during Roosevelt's corporate, National-Socialist Raw Deal in the thirties, suicide is illegal because it destroys property, which belongs to the State, (read: Federal Reserve Corporation, a European company). I'm not kidding. The UCC oversaw the growth of the most extensive gulag archipelago, in man's history—the US prison system—and the majority of people, who are in prison in the US, were convicted for non-crimes. Most of them are people of color, so technically it's a UCC crime not to be Europoid in America. Institutional racism's most glaring example is the majority population of every jail and prison in America, which is predominantly black and Hispanic. This handily reinforces the What To Think Network's representation that only Europoids 'obey the law.'

All of the genuine criminals, including the clear majority—the ones in government and government-sponsored businesses—need to be held accountable now, don't you agree? Does anyone besides the Depression Babies, still wink at their crimes? This treasonous Congress has passed legislation that is so openly oppressive, that the executive and judicial branches are going to have to put up or shut up. 'Put up', means more and more federal, fake-terrorist mass murder events, like the WTC bombing, in order to make the PJ folks suitably paranoid; 'shut up', means they need to go to jail and potters' fields after treason trials, and make way for genuine, minimal government. What do you reckon the more productive choice is?

Even if the Corporation imported the entire Chinese and

Russian armed forces to sufficiently swell the Homeland Security Abominations' ranks, in the event of overt martial law, the Chinese generals, who are clearly in charge of that new Axis, know Sun Tzu's writings by heart, even if the US politicos can't comprehend "THE ART OF WAR", and you can bet they'd quickly regret trying to conquer such a vast, still-armed populace.

The Germans under Hitler, the Russians under Stalin and the Chinese under Mao preferred 'safety' and relative prosperity to freedom, obviously, and they didn't mind that a few million innocents were rounded up and shot, as long as they were personally left alone and prospered. I believe that even the American Pajama People would violently object to seeing truckloads of people taken from their homes in the wee hours, and hauled off to execution chambers and work camps, even if only a few of the victims were Europoids. The Germans, Russians and Chinese mainly wanted to conform, so that they could continue to enjoy the alleged fruits of socialism. This is how Pajama People are, when tyranny is allowed to prevail, of course. None of those people had any particularly strong traditions of personal freedom, which is probably why they so eagerly gave up their firearms.

If the chemtrail genocide agenda and the death tower agenda had been successful the Corporation might have had a fair shot at global tyranny, but they knew that they lost their initiative when most Americans, even the PJ folks, refused to condone the Waco massacre on cue.

According to the Georgia Guidestones, on which the world order's global genocide and dictatorship agenda is clearly and artfully inscribed in several languages, the world's population was supposed to have been reduced to a half billion, before January 1st, 2000. Carol and I read that 'prediction', when we gifted that remote, extravagant, granite monument, three years ago.

The rest of that inscribed plan reads like an Orwellian nightmare, and is apparently a franker expression of the United Nations' true agenda. Instead of traveling to that obscure spot, an hour and a half from Augusta, Georgia on secondary roads, you can read the slabs' inscribed agenda on www.georgiaguidestones.com. You can disregard the fundamentalist Christian overtones on that site, of course, because the inscriptions speak for themselves: we were supposed to all be dead before 2000.

If you were told that the Lucifer Trust publishes all of the United Nations' official literature, you'd probably consider the UN to be less than forthcoming about their hidden ideological agenda. In fact, Alice Bailey, the infantile satanist who began claiming that Krishnamurti was 'The Return of Christ' in the 1920s, later volunteered or agreed to set up the Lucifer Trust, to publish the United Nations' literature. Shortly after World War II, this was reported in the press, and the name of the trust was quietly changed to 'Lucis Trust'. Alice Bailey wasn't mentioned again, in the context of the United Nations, as far as we know.

Lucis Trust publishes all of the United Nations' official literature and Theosophy was to have been the world's fake global religion, before now, if the Georgia Guidestones' chiseled wishes had been granted.

I guess 'Convert or Die,' would have been the rallying cry of their religion, in that case. Now, they're an assortment of schizophrenics, who like to pretend that they're God.

If you've got the time and interest, read Theosophy's nauseating, spirit-crushing ideology and see how closely it mirrors the ancient Vril 'dark sun' ideology. For the folks on the fringe of Theosophy—the starry-eyed new-agers—a slightly altered ideology similar to dualism is represented, in which the world is seen as bad and 'ascension', or rescue from this life, is seen as the ultimate goal. You'll probably stop thinking of Nordic blondes as stupid and comical, after you read some of this creepy 'dark sun' stuff. I was married to a blonde for 22 years and she was anything but stupid or comical.

Did you know that Rudolf Steiner left Theosophy in disgust, after Alice Bailey offered to represent him as 'the Return of John the Baptist'? He had objected to her scamming people about Krishnamurti, on ethical grounds, and she had assumed that Steiner was just jealous. Nothing good ever came out of Theosophy, but Steiner's work is an excellent source of insights and information, if you can get past his personal assumption of arbitrary authority. D Bradley spent some time with Krishnamurti, before the old man died, and said that he was a genuinely sweet guy. He kind of reminds me of Emperor Hirohito, who was similarly used by the Japanese occult/corporate hierarchy, but a difference is that the emperor eventually acted against the old world order and stopped World War Two, but Krishnamurti never actually denied that he was Jesus Christ.

Is it clear by now that Japan was never a real threat to the American economy? SONY stands for Standard Oil of New York.

Fascism is functionally, if not ideologically, the unfair combination of business and government, that's designed to exploit and reduce the world's population and generally rob us all of our birthrights, including breathing. Communism is fascism's butt-ugly fraternal twin, spawned by the banker trolls' adulterous coupling with Russia and China, through 'The London School of Economics.'

Tweedle Dee and Tweedle Dum. Both ideologies are rooted in the excessive centralization of political and economic power, and the attendant, bloated bureaucracies. The horror of the New York Stock Exchange is made possible by fascism, and it's essentially no different from communism, because it favors the few who pirated the world's natural and human resources. Dissolving centralized power will put all businessmen on an equal footing, at last, and state-enabled monopolies will no longer be possible.

The What To Think Network, especially National Public Radio, might have conned you into believing that communism is 'left wing', and that fascism is 'right wing.' To confuse you even more, NPR and your Marxist college instructors call Russian communists and American fascists, (Republicans), 'conservative,' and rabid National-Socialists in America, (the hippie generation's legacy), and anti-communists in Russia, 'liberals.' Hopefully, you began to educate yourself after you earned your college degree and have put down your bong forever.

Technically, I suppose communism and fascism are the same wing, and the other wing would be anarchy but really, ideology has nothing at all to do with government, any more than it has a part in the internet's workings or your relationship with your siblings, because the simple truth is that we need to have absolutely as little government as possible, so that we can have as much freedom and personal accountability as possible. This is just an idea whose time has come; I'm stating the obvious again. A region in northern Spain is run on a sort of communist system, but it's a grassroots network of Basques, who decided to pool their resources and reduce the significance of personal ownership. Really, it's based on their Celtic traditions, so they're comfortable with that. A community or small region in America might be run on entirely different social principles, but who cares if everyone agrees to the system and has a voice in it? To apply any ideology to a central government is to invite corruption and tyranny. Central governments have to be as unobtrusive as the guys who operate the internet are. The internet works extremely well, as anyone can see, so why can't government do the same now? We're ready for that.

It's past time for that cancer-ridden beast, heavily-centralized political and economic power, to expire, and orgonite is the magic bullet that's being used to perform the coup de gras on this behemoth and to enable the flowering and fruition of a fresher, more liberating government paradigm.

The foundation of all physical activity is etheric activity, of course. We gifters, and predator blasters, have been working in the ethers to create the right conditions for the unfoldment of a long awaited human, but spiritual civilization. Nothing can happen in the physical world, unless preparation is made in the etheric matrix, of course. The bad guys knew this right along, which is why they put such a premium on polluting, then pirating the earth's natural energy field and stealing its resources. We're learning the principles of energy healing now, and it's not too late to close the wounds that these beastly misogynists have increasingly inflicted on our world, for millennia.

The internet spontaneously appeared, developed and spread throughout the world, without any help or effective interference from governments or government-sanctioned corporations. I think Russia and China are the only countries left, where the inhabitants are so heavily monitored and intimidated by overt electronic surveillance, that they rarely seek out empowering information online. We, in the West, are just as heavily monitored by quasi-government secret police agencies, but these dispirited, leaderless cabals, in the West, are a lot weaker than Russia's and China's, so they won't dare enforce their horrifying, draconian laws, in most cases. They know we're all armed, at least in Switzerland, Brazil, Punjab, Uganda, the US and several other countries.

I've published a book that strongly advises the dissolution of the treasonous US Government, as soon as possible. How many of their new treasonous laws do you reckon I've just broken? How many of those fake laws are you breaking, by buying and reading this? They've made it impossible for me to wire money to our friends in Africa any more, because my name is on an exclusive list of 50,000 Americans who are being officially interfered with, in annoying ways, by the Homeland Security Abomination.

Three years ago the FBJ and the Moscow, Idaho police department were telling people who know us that, 'Don Croft is Idaho's Number One Terrorist.' The executive assistant to the commander of the Idaho State Police is an ardent gifter. She told us that there is no official 'terrorist list' in Idaho, so I guess those feds were just saying it for effect. I still feel proud that they'd give me that much credit. I certainly wish to personally and lawfully terrorize all of those traitors and their corporate masters, because that's a public service and I do love to serve my fellows.

Now and then, I get an email from new friends reporting that when they got in touch with me, their computers were crashed by hackers. A fearful person would take this as an effective threat to 'stay away from Croft', but these folks are, rather usually, inspired with fresh resolve and they generally spend more effort busting up death towers, on account of that confirming experience. As a rule, if the bad guys don't express their displeasure to us, individually, it means that we're just not hurting their treasonous agenda enough. The best fun starts when you turn the tables on your tormentors and make them afraid of you. We do this by sending them life force, through our hearts... with a lot of force.

You'll be far less vulnerable to the assassin agencies' hackers, if you stop using email software. Do your mail online, as we do, and it will be nearly impossible for them to crash your computer. I know it's a little less convenient, but it's more convenient than having to replace your hard drive, don't you think? If you're using an American internet provider, the NSA will make sure that the provider won't let you use foreign email software, like Phoenix. The feds want you to use email software that's loaded with 'Bill' Gates, so that their hackers can get into your hard drive more easily, to destroy it. I bet there's not a single American internet provider, who won't bend over for the NSA. That's why I chose a reputable Montreal server to host our two websites: www.ethericwarriors.com and www.worldwithoutparasites.com.

It's generally assumed, by people in the gifting network, that if we're not under occasional assault by government-payrolled hackers and predatory secret police psychics, poison-mongers, character assassins, burglars and electronic weaponry, it means we're not hurting their masters, so we get busier and work smarter, until we start getting those little confirmations again. We do get tired of it, though. We're not masochists and I can tell you it's useful to have a good watchdog in the house. Before long, this world order will be defeated, then we can all concentrate our energies on turning this incredible planet into a garden.

Ten years ago, their beam weapons, along with their aerosol, powdered, injected poisons and toxic metals, might have killed us all, but the new technology we've developed counteracts all of that and enables us to turn the tables on these predators. Most of the psychics in the West, who have personal integrity, were murdered on the sly this way, before five years ago, which is apparently when the new army of trained Monarch and Tavistock psychic predators were fully deployed. The majority of skilled psychics work for the assassination agencies, as intelligence gatherers and predators. We're damn lucky to have found some surviving good ones, in fact. As this network spreads in Africa, we'll find new friends and coworkers among the powerful but benevolent psychics there, and I think we'll be home free, because these folks are heavy hitters, compared to the insipid, Europoid and Asian dirty-magic practitioners, whom the world order have been relying on.

The rest of the world's national regimes lack the money and infrastructure to monitor and harass their citizenry. Even going to Canada for a few days gets us some welcome relief from the incessant assaults, because the Canadian government isn't as keen on destroying its own citizenry as the US Government is. Their hearts just don't seem to be in it, unless they're going after empowered Quebecois folks in the network, like Steeve DeBellefuille, Jacques Lasselle, Denis Couture and Alexandre Emarde. Steeve set up Etheric Warriors for us and connected us to his server, and Jacques is administering it now. We gather that MI6 sewer rats are the ones mainly bothering the French guys; not the Canadian secret police.

I've been told that there are anti-government riots in China, these days, so maybe tyranny will end there pretty soon. Russia's been on the point of anarchy for over a decade, but somehow the Russian military and naval forces kept growing and consolidating—certainly financed by European and American bankers and their laundered dope money, since there's no lawful prosperity in Russia any more. A Russian friend of Dr von Peters', who was the personal physician of one of the previous Premieres in commie days, had to come to America to earn a living, as a heart specialist in recent years.

All of the soldiers, police and assassins on the planet, can't block the path of a billion Chinese people, who simply choose not to be oppressed any more, so I'm not worried about the Chinese. They came pretty close to waking up, after their children were murdered en masse in Tienanmen Square. See how the Waco massacre mirrored that one? There are no coincidences at that level: National-Socialism appeared in America the same time it appeared in Germany, for instance, and China soon followed. Russia did a flip flop from communism to fascism. The government, is now openly run by Russian corporate magnates. So much for 'London' Lenin's promise to hang capitalists with the rope that they, themselves bought.

Brainwashing was pioneered in ancient China, so we shouldn't expect that regime to fall any faster than the Bush Senior regime is falling, I suppose. Maybe Japanese gifters will flood China with orgonite, on 'business trips', in coming days. I have a hunch that Japan will soon be the first country where orgonite reaches public awareness, perhaps due to my Japanese brother, Tetsuzi Moriwake's, pioneering, untiring promotional efforts, in Hiroshima and elsewhere.

We predator blasters and intel gatherers are fully convinced that Beijing calls the shots for the world order now, based on several reputable psychics' consistent impressions during the past eight months.

'MalWart;' 'nuff said.

This is exactly how London undermined competitive nations, too: heavily subsidized industry aimed at undermining foreign economies with cheap, crummy, imported, manufactured goods. In MalWart's case, it's pretty obvious that the US Government is also subsidizing some of those cheap wares, in an effort to further obliterate the independent, small-merchant class.

I want you to keep an open mind about this. See whether more and more corroborative evidence of China's recent acquisition of the US Government shows up during the coming months, before you make a judgment about our observations.

If you feel a need to play 'devil's advocate', every time you encounter a view that's contrary to The What To Think Network's party line, it just means that you've been effectively programmed to police your own thoughts. What else do you suppose the What To Think Network is set up to advocate except 'the devil,' after all? Why, in fact, do you think that nobody rushes to play 'angel's advocate', every time the What To Think Network rattles on and on about 'the terrorist threat,' non-conspiracy theory and the host of other standard misrepresentations? Nobody thinks to do that because watching TV, even the 'educational channels,' intentionally hypnotizes and stupefies people.

Righteous anger can get us moving in a positive direction, but impotent rage stops us in our tracks and ultimately serves the hidden, walking parasites' enervating purpose. Most of the time, when one experiences an activation of one's sense of destiny, the first emotion is anger, and that's probably in order, because it usually comes from a realization of how thoroughly we've been exploited and held back, all of our lives.

When one finally discovers that his or her entire life up until then, was largely directed by cynical others, what other emotion would you expect to surface first? A balanced person experiences anger; a dissociative-programmed, indolent Pajama Person feels not much of anything, and grabs another beer or lights a fattie.

It might be helpful to recognize that exposing 'Them' is more effective at stopping their generations-old mass hypnosis and environmental piracy, than shooting them would be. These faceless corporate vampires have no more claim on the earth's resources and occupants than the unwashed, fratricidal Kings of Spain, France and England had lawful claim on the Americas and countless millions of autonomous inhabitants.

Corporations, like European monarchies, only exist on paper, after all. Let's just discorporate the parasitic ones, including the felonious Fed and the pungent skid-mark, that calls itself the House of Windsor, and enable prosperity for anyone who wants to earn it by honest means.

In the absence of centralized tyranny, nobody in the world is going to starve, I believe, because locally-operated charities will more likely hold their recipients accountable in the short term, so most of them will choose to earn a livelihood in appropriate ways. National-Socialist welfare agencies simply use the disenfranchised as political chattel, so it's not in their interest to actually help them become self-subsistent. Most people want to contribute to the common weal. National-Socialist welfare distorts that natural urge and keeps families apart, by encouraging moms to be breeders for the State. They have to keep making babies in order to get their monthly checks up to the poverty line and if the social worker catches a daddy in the house, the mom and her brood are cast out into the street.

The dole is how Hillary Clinton got into the White House, by the way—hubbie's handout. I bet Vince Foster was ultimately sorry that he helped her ersatz 'career.' Why not just put Lizzy Borden in the Oval Office, while we're at it? American national politics, since the Civil War, is as spellbinding as 'Punch and Judy', or "THE LITTLE SHOP OF HORRORS". Under the emerging paradigm, the office of the President would be little more significant than a notary public's.

Most people are naturally charitable these days, in case you didn't notice. Poor people are usually the most self-sacrificing when it comes to helping others.

By the time we left Spokane in September 2005, I had a clear picture in my mind of how our cetaceans project would likely unfold here, in Florida. None of that has come to pass—it's happening in a more interesting, gradual, organic and educational way, and Carol's no longer laughing up her sleeve at my naïve, grand assumptions.

I assumed that by now, for instance, there would be people along all the US coasts and beyond, handing orgonite to dolphins and receiving love and information from them in return. All that will likely happen a little later on, throughout the world. I sometimes tend to jump the gun, but that's okay. For instance, I assume the US Government will fall within a couple of years, but maybe it will take three or four years for tyranny to end here.

On the other hand, I naively but correctly assumed that our wonderful African gifting cohorts in Western Kenya would likely meet some dolphins, if they could only take some orgonite out to sea, near Mombasa. I appealed to the generous Western network for funding their trip, collected it for them, and they traveled across the country in comfort, and had some wonderful, dramatic interactions with a couple of pods of dolphins, who swarmed the boats on successive days, to get the orgonite they tossed over the side. David Ochieng Omolo's first hand report of that adventure is on www.ethericwarriors.com.

Carol and I had our hearts set on getting a catamaran that's seaworthy enough to cross the ocean and big enough to live on, but after traveling the state for the first month here, checking out the available boats and talking to brokers and sellers, we suddenly realized that all we really need, for now, is a fast motorboat that's seaworthy and can carry a few people out to sea and back in an afternoon, so we got the twenty-three foot Zodiac Pro with a 150hp motor and we don't even get our feet wet, even in rough seas.

It was another month between the time we put the money down on the boat and got it in our yard. Meanwhile, Jeff McKinley, South Florida's premier orgonite tosser, Carol and I occasionally swam out, past the surf at two local beaches, to drop loads of orgonite for the dolphins to collect and disseminate out there.

After awhile, we simply went to the nearest beach, in the late afternoons and Jeff and I spud-gunned orgonite towerbusters out, past the surf for them. Carol walked the beach and communed with the nearby dolphins then, which we hadn't seen yet, but whose presence we all felt.

We discovered why the dolphins were unable to approach the shore, when we were there, apparently. We found and disabled a line of underwater, deadly-orgone-radiation transmitters, (presumably the US Navy's local, anti-dolphin sonar weaponry), north of Jupiter Inlet, five miles from shore.

Jeff's got quite a bit of latent psi talent at this stage, and some of that has been waking up just from being around Carol, on our humble sorties. He's also been developing some interesting new, interactive devices, which may play a role in stopping the upcoming HAARPicane strategies. These devices interface with huge undines and other friendly elementals, as I mentioned earlier. Other psychics have confirmed that, during the sessions.

These new and varied devices are a little like Aladdin's lamp, and he's going to deploy them, before long, to appropriate individuals here and abroad, who can work together to toss energy at any artificial storms that manage to sneak past our regional HAARP-disabling efforts on land and sea. I think he's the first to create devices that apparently interface with massive earth elementals. This 'new awareness and opportunity' stuff gets pretty heady, and I want as many people as possible to experience it.

Three weeks after we arrived from the Pacific Northwest, HAARPicane Wilma arrived at our small, beachfront town, from across the Florida Peninsula, as I mentioned before. When it reached the ocean, downwind, it quickly dissipated, probably as a result of Jeff's towerbusting efficiency with the coastal HAARP arrays. That's the opposite of how a natural hurricane behaves and natural storms never travel in a straight line, either, as that storm did, as it crossed the state from Naples to Jupiter.

Seventy miles south of ground zero, in Ft. Lauderdale, property damage from the storm was pretty severe, due to stronger winds, apparently because nobody's been systematically disabling the HAARP network and death towers there yet. The really big HAARP arrays are usually 'hidden in plain sight', in larger cities, where the PJ folks are most soundly asleep. At ground zero, (our town, north of Palm Beach), most of the large, screened enclosures that are popular in Florida weren't even damaged, thanks to the reduced wind speed of the storm as it approached us. Thanks, Jeff!

Carol had gone back to the Northwest to take care of some personal business at the time, so I experienced the storm by myself, much of it sitting in the car and doing my email, via a little inverter, plugged into the cigarette lighter. The feds turned the power off in South Florida, to create the impression that the storm did a lot of damage. They forgot to turn off the phones' landlines, which use the same power poles—duh! Predators are cunning, but they're not very intelligent.

Newsreaders lie a lot. In the first part of the storm, local reporters on radio were giving accurate accounts from listeners, who were calling in their observations, but by the time the eye of the storm arrived, the What To Think Network dismissed all the local reporters and superimposed their own professional liars from up north, who gave bizarre, emotional accounts of damage that never occurred. That was all for national consumption, of course, and they didn't want local folks to network on the air. One would think that Pravda's imaginitive Sorcha Faal was in the What To Think Network's driver seat, in South Florida that day. This freak show was pre-planned, because all those national media folks hit the ground running, when the eye of the storm arrived. It reminds me of how Bush Senior and Bonehead Clinton set up the 'New Orleans Hurricane and Flood Relief' fund, several months before the feds blew up those levees, in New Orleans.

That tropical storm was obviously supposed to be hell on earth, because the Homeland Security Abominations who had us under box surveillance since we arrived in Florida, fled a couple of days before the storm, and they came back in full force the next day. It was a nice little break from their omnipresence.

I went through a real hurricane—ten years before in a small, open sailboat in the Gulf of Mexico near Yucatan. That one seemed to be more natural, but I doubt that there have been any natural hurricanes since the 1970s, or perhaps since Babylonian times. For several days before and after the eye passed, the sky was full of very dark, DOR-laden roiling clouds, and it was very windy and rainy. This time, the sky was cerulean blue and filled with lovely Sylphs, a few hours before the eye arrived; it was entirely clear, calm and the ambience was invigorating, within an hour after the eye passed. The clouds, throughout the storm, were light, not dark. It rained for about two hours before and after the eye was here, but it was calm and the skies were brilliant blue, before and after that short event.

The calm period in the middle lasted about forty-five minutes. That storm was wishfully predicted to be cataclysmic, but the wind here, didn't get over seventy miles per hour, except for a few gusts at the back end of the eye. The target, (North Palm Beach County, which is where Jupiter is), was also predicted a week earlier, though at that time the storm was moving away from us, on an erratic path in the Gulf of Mexico, far to the southeast. I don't think there's a distinction between 'prediction' and 'direction', when it comes to violent storms reported on the What To Think Network. It may be that announcing the storm's precise, eventual target was an open joke, at the expense of the Pajama People's simpleton mentality. The bad guys, too, have a sense of humor and sometimes their jokes are pretty good.

In addition to Jeff's previous, local HAARP-busting effort, several friends in the US and abroad were directing energy at the storm, in order to pressurize the eye. This activity is sort of like prayer, also sort of like an answer to prayer. This new era that humanity is entering has some curious new possibilities. The bad guys are openly predicting a devastating HAARPicane season this year, for the region north of us, but we know, by now, that it will be a lot milder than the previous one, thanks to this network. The previous one was a great deal milder than predicted, too, and so on, for the past five years since orgonite began to proliferate.

Carol bought a barometer last summer, when we were in Spokane, and I assumed the thing didn't work, until the needle suddenly dropped to the bottom of the scale, shortly before the eye of the storm arrived, then jumped right back up as soon as it passed. I suspect that having the cloudbuster nearby effectively keeps the atmospheric pressure high, regardless of humidity. That may be one of the stronger evidences that an orgonite cloudbuster is affecting the atmosphere, also why no, (natural), violent storms or strong winds have occurred near anyone's orgonite cloudbuster.

Another proof may be the obvious ionization effect that orgonite has on the environment. A healthy organism and atmosphere is characterized by over-abundance of electrons, (negative ions), so we've been assuming that ionization is simply an expression of positive orgone radiation. A deficit of negative ions in the atmosphere results in smog, lightning, strong wind and human misery; a deficit of the same, within the human body, results in disease and parasites.

Deadly orgone radiation's net abundance in the atmosphere is characterized by a deficit of negative ions; a positive static charge. In this environment, toxic solids are colloidally suspended in the atmosphere. That brownish yellow haze is smog. Bringing a bucketful of orgonite towerbusters in your car into that energy field, generates a lot of capacitance inside the vehicle, so when one contacts the ground while touching the body of the car a sudden surge of electric current is enabled, as the circuit from the metal body of the car to ground is completed. Rubber tires apparently insulate well enough that the energy of the car isn't constantly grounded. This is an approximate explanation and I'm not convinced that anyone can adequately understand, let alone describe, the dynamics of electricity.

This is most easily experienced by carrying a lot of orgonite in your car to an energy-poisoned area, such as the downtown of a large city, that hasn't been gifted yet. Throughout the world's population centers, the death towers are distributed in a ration of one for every two thousand or so people. Just about every building you find that is over three stories tall, has the death transmitter panels and/or rods, around the top storey and on the roof, so you can imagine how densely concentrated the deadly technology is, in every downtown area.

When you put your foot on the ground, while your hand touches the car's metal body under those conditions, you get a savage, static jolt, even on a warm, humid day.

When Jeff started gifting a particularly smoggy city, in North Carolina, years ago, so much energy went through him, as he stepped onto the pavement that his trunk flew open, and his car alarm went off.

I, too, experienced that unpleasant shock few times, during urban gifting sorties. It takes a little while for the orgonite to ionize the atmosphere and dissipate the smog, but it always happens fast enough for you to track it, so keep watching. Don't just gift and run, if you want to get your well-earned, requisite visual and improved-ambience confirmations.

Psychics and energy sensitives don't need these sledge blows of physical sensation, to get confirmations for the energy dynamics of orgonite. 'Good for them!' I say.

Jeff was at the local beach recently, and he lay down and focused on asking the dolphins to show up. He fell asleep in the process, but the excitement of the other people on the beach soon woke him up, when two dolphin pods showed up. Spinner and bottlenose dolphins started leaping into the air, in the shallow water, just beyond the surf. The bottlenose dolphins hadn't been seen in the area for several years, and we'd only been seeing spinner dolphins, since we started gifting in the local seas and waterways, in recent months. Before, the spinner dolphins had showed up in that same spot, a day after Jeff and I took a load of orgonite out beyond the surf. We did that the day after Carol and I first arrived in the area.

While the boat's been in the shop for the past month, lots of spinner dolphins have been seen along the beaches for ten miles or so, north and south of Jupiter Inlet, which is where we go out to sea, on our local gifting sorties.

Chapter Five

Exoskeleton

Orgonite is entirely harmless. If that weren't so, there would be no grassroots, global, gifting network; this essential, planetary healing work would be left in the hands of a few talented folks, as was the case with Dr Reich's powerful, but potentially dangerous cloudbusters. 'A few talented folks' would never get it done, because winning any war is a numbers game, and the bad guys' genocide wishes might have been granted, in that case.

> 'Gifting' is what we call distributing simple, orgonite devices, (usually homemade), in the environment in a systematic way, to initiate and sustain positive energy transformations. 'Orgonite' is the energy-producing material that is basically made up of resin and machined-metal waste.

This simple, powerful material was eventually developed from the earlier findings of Dr Wilhelm Reich, who called the trans-dimensional, subtle, energy matrix of the universe, 'orgone.' He invented that term after scientifically discovering that the energy released during sexual orgasm is pure, creative force. We like to honor his life of education, healing, ground-breaking research and selfless service, by using his term for the energy matrix, which others have named ch'i, prana, life force, animal magnetism, odic force, vril (!), etcetera.

Until we all came along, the What To Think Network had essentially erased Dr Reich from public memory, a process that was cemented within a decade or so, after his death, when crackpots and scholarly pedants, rather than serious researchers, had felt obliged to carry his banner. These are among the sorts of people he detested, when he was alive, strange to tell. The other class of villains he named, besides sycophants, fake scientists and mystics, were corporate business types. I think the only mystics, who were popular the West in those days, were Theosophists, so who can blame him? Now, his work and writings are being resurrected, due to renewed public interest in orgone, and that's very gratifying to many of us.

Dr Reich was a sweet natured, unshackled, self-effacing genius and productive scientist, who was able show us how to consistently and systematically apply the creative force of healthy orgone to achieve widespread, beneficial results. When the What To Think Network was taking him down, in the early 1950s, after he achieved wide acclaim for improving the weather and curing cancer, they tried to make him look like a sexual deviant, but he publicly recommended monogamous heterosexuality, as the only healthy expression of this essential human drive.

In fact, when he was the favored protégé of Sigmund Freud, in the late 1920s, he was expelled from the communist party, after discovering that the 'angry workers', who attended the rallies were that way because they weren't having sex with their wives any more. After counseling many of them to resume sexual relations with their spouses, they just didn't feel angry any more, and so lost interest in supporting communism.

For that matter, Freud blackballed him in the 1930s, because Reich was successful at healing mental illness; Freud was only concerned with analyzing it, as you probably know. Freud's analytical skill found a proper outlet, finally, when he headed British Secret Service's 'Tavistock Institute, the world's premier brainwashing institution, in the late 1930s. Fortunately for us, Dr Reich was driven out of Europe and the psychiatric profession, and then put his first efforts into researching the characteristics of orgone and applying his knowledge in the US.

The day before he was murdered by poison in a federal prison, in 1957, Dr Reich handed a manuscript to his sixteen-year-old son, Peter, who is still around, apparently. The book allegedly has a full description of how gravity works. That was a day before he was to have been released from prison. He was railroaded through a kangaroo federal court, on account of his healing successes. There have been many, many casualties in the Federal War on Healers, since the mid-1930s, but those casualties have always been on our side until now. Before the American courts were handed over to a London corporation, by FDR and a treasonous Congress, the American Medical Association simply hired the Mafia to dispatch genuine pioneers, in the healing arts and sciences.

The spontaneous, grassroots campaign to heal the world and end tyranny with orgonite distribution and orgone-tossing, (predator safaris), is indeed, a numbers game, not a talent show. The arena where the more gifted few among the network can shine, is in the creation and intelligent use of potent interactive devices, which usually include orgonite. Reputable, skilled psychics can gather intel for our mass-murderer hunts, and can accurately analyze the effects of new inventions and perhaps help find applications for them.

The prize takers are the ones among us who simply distribute the most orgonite, because every war's object is to take and hold territory. Disabling the death energy matrix in any territory takes real estate away from the occult/corporate pirates, who polluted it in the first place. It's much the same way we get rid of parasites in the body with micro-current, when we use a zapper: ionize the parasites' immediate environment, and they can't stay there and exploit it any more. Blasting key predators in occult/corporate hierarchies is like destroying

the parasites in the body with micro-current.
The emergence of the proprietary approach to healing with orgonite might be a matter of timing, and the time may be near when people who have special energy-related talents can network properly and internationally, on a very wide scale, thanks to the internet. We're just the vanguard, at the moment.

In the formative years, most of the people in the network, who were promoting proprietary devices, were disreputable, but in recent years most of the inventors have been team players and self-effacing, (meek). The most dramatic work will get done by masses of committed, ordinary folks, distributing simple orgonite devices to other people, after we're done disabling the millions of death transmitters on land and on the seabed.

This summer, (2006), we intend to have a few proprietary devices in use, conceived and artfully crafted by Jeff McKinley, to have on hand to suppress any HAARPicanes that get generated, in spite of our best efforts to disable the weather warfare facilities. Dave Emmett in Barbados has been instrumental in getting it done, in the Caribbean region, beyond the US.

The genuine aficionados are pretty spread out: Kelly McKennon and Ryan McGinty in Washington State, Cesco Soggiu in Iceland, Tetsuzi Moriwake in Japan, Jeff McKinley right here in our little town, Carol, D Bradley in LA, Tracey Ann in UK, Eric Nagal in the Phillipines, Kizira Ibrahim in Uganda, and vo Joanna in Brazil to name a few and many more gifted, inventive, cordial and committed people will probably show up, before long.

Cesco, an accomplished, young multi-media artist, as well as inventor, developed a unique, three-dimensional coil form, which many of us have been using to enhance our interactive devices. His site is www.soulbalm.com. Carol and I feel that small versions of his coil were instrumental in suddenly stopping the seismic activity at the Yellowstone Caldera, when we buried twenty-six earthpipes, (most of those had the coils in the orgonite component), around the hundred-mile perimeter of that caldera, in October 2004. A massive eruption was imminent, then, which threatened to erase human civilization.

The September day after we left our first batch of orgonite, (made by Jeff—we were staying at a campground, when we arrived in Florida last fall), just beyond the surf in the sea near us, some surfers at that beach saw some black-backed spinner dolphins swimming very close to shore, in the spot where we'd dropped the orgonite. The local pods are bottlenose dolphins and they're grey-tan colored, so the spinners apparently came from somewhere else. These smaller, black-backed dolphins got the name because of the way they like to spin, as they leap out of the water.

A notoriously friendly member of the bottlenose dolphin pod had been murdered in Jupiter Inlet, some years before, and the pod wasn't seen until Jeff recently had the experience on the beach, which I mentioned in the previous chapter. Carol recently saw spinner dolphins in the Loxahatchee River, a mile and a half inland from Jupiter Inlet. She was driving over a bridge at the time. Tossing orgonite into the water has turned the inlet blue-green again and full of fish, after many years of being polluted and lifeless.

Our first direct encounter with the spinner dolphins happened in mid-December, during our first gifting sortie out of Jupiter Inlet, which is when we found and disabled the dozen or so new underwater transmitters, directly to the northeast. In fact, the pod was hanging out around the last underwater tower we gifted that day, about five miles out. We ran out of orgonite there, (we later returned on New Year's Eve to disable the few remaining towers beyond that point), but Carol tossed them a couple of 'dolphin balls,' which she makes especially for dolphins and we keep on board for these occasions. Several reputable orgonite vendors, including:

www.ctbusters.com,
www.quebecorgone.com,
www.orgoneaustralia.com.au,
www.orgonise-africa.net,

make and sell their own excellent versions.

Steve Baron and crew, in Toronto, meticulously made and distributed thousands of dolphin gifts to people around the world for a year, before we got started here, in Florida. I think that generated a lot of momentum for this project. Steve generously sends us hundreds of those at a time, which we deliver to the cetaceans and share with other gifters. Steve also enabled Georg Ritschl of www.orgonise-africa.net, to gift along the entire Indian Ocean coast up to Zanzibar.

It may be that the same pod, which has been showing up at the end of most of our local gifting sorties, also greeted us once near Key West, where they specifically asked Carol to lay a line of orgonite along the shore, to block the radiation of a 'weather ball', at Key West Naval Air Station. We did that on our way back from gifting the sixty miles of reef, that day. If you ever come across a radome on your sorties, please gift it! They're all particularly heinous components of HAARP weaponry, in addition to whatever legitimate radar function they might also serve. Surprisingly, it only takes a couple of towerbusters to neutralize one of these horrors, if you can get within a quarter mile or so, otherwise you'll need to toss a lot more towerbusters out farther away, hopefully surrounding the facility. The most dramatic effect we got from disabling a weather ball, was when we erased a creepy looking, dark conical field around one near Spokane, Washington, with a couple of towerbusters. The entire atmosphere lightened up, for several miles around, right away. It's amazing how much bad energy these domes can throw out.

Carol's gone to Hawaii a couple of times to swim with the spinner dolphins, which frequent a bay on the Kona coast of the Big Island. If you want to meet dolphins, you don't need to pay for a tour. You'll get better interaction, (especially if you take them some orgonite), any day and for no charge, if you just swim out into Keelakaua Bay. That's where Captain Cook first landed in the Hawaiian Islands.

Rockefeller-funded environmentalists, (Nature's Conservancy), are trying make it unlawful for you to swim with those particular dolphins, but I don't think they'll succeed. Short of that, they'll probably continue to try to scare people away from those wonderful healers. They go out in boats to terrorize swimmers—no kidding! Strange world, eh? Just ignore the environmentalists, when you go there.

One of those environmentalists swam with the dolphins on the sly, when the rest weren't watching, and Carol encountered her on the beach on her way to swim with them one day. The woman was weeping uncontrollably and told Carol, 'I had no idea!'

The Florida spinner dolphins show up for us in odd places, including the shallow Indian River, behind Hutchinson Island, (about thirty miles north of here), when we laid some orgonite along that waterway from the south. Later, when Jeff and I laid a line of towerbusters for forty miles from Jupiter Inlet, past Palm Beach to Boca Raton and back up the intra-coastal waterway toward Jupiter, the spinner dolphins showed up as the sun was setting, and we'd just tossed our last towerbuster—that was in a narrow part of the waterway, a few miles south of Jupiter, on our return leg.

That time, they let us approach much closer and Jeff and got a couple of Carol's dolphin balls ready for them, as I slowly drive the boat in a circle. As per Carol's instructions, when we saw the dorsal fins we immediately slowed the boat and turned toward them. When they changed direction we steered the boat in a slow, tight circle and they paralleled us about thirty yards away. Right after the second ball was tossed, we saw an upwelling of water, right next to the boat, as from the flip of a dolphin's tail.

Carol said that this pod has never been close to people, and that it may be a month or two before they'll invite us to swim with them.

The pod that our African cohorts recently encountered near Mombasa were a little shy, as David Ochieng noted in his report, which is on http://www.ethericwarriors.com/main/index.php?page=forums&action=topic&t=180, but I bet they'll be able to swim with this pod, when they return, because they suddenly appeared and swam all around the boat quite close, after Mrs Odondi felt inspired to toss out some orgonite.

David and the group's local guide, ('beach boy'), got into the water with the dolphins the first day out, but the pod vanished after David lost his balance for a bit, and made a splash. I suspect that their next trip to Mombasa will include many more people, and I hope to get them a digital camera for a graphic record. Judy Lubulwa and the other new gifters in Nairobi may want to go along with our Western Kenya cohorts, next time.

I love the way Africans can network, and Kenya may soon

be the showplace for gifting's power to heal the environment and society. Uganda, where the first Africans took to gifting and cloudbusters, is already so nice that it's harder to track the improvement there, but Kenya is in dire straits, so the observed results of gifting are more dramatic there.

When Carol's grown kids, Nick and Jenny, came from Idaho to visit us in early February, Carol dragged the tailored boat down to Key West, and took them out to Sandy Key, which was the terminus of our 40-mile reef-gifting run a couple of weeks before, when we encountered the spinner dolphins.

Sandy Key is the place where huge sailing catamarans take groups of snorkelers from Key West to tour the reef, several times a day. It is six miles west of the island. Cosgrove Shoals, another fifteen miles west along the reef from Sandy Key, is where people normally go from Key West to swim with bottlenose dolphins.

Another place nearby is Bimini, which is directly across the Gulf Stream from Miami, about sixty miles. That dolphin pod wasn't seen for an entire year, before Steve Baron showed up there, last year, and tossed some of his dolphin gifts in the area, then they showed up immediately, eager for contact and perhaps more orgonite. They probably knew he was coming, before he knew he was going—we can't begin to fathom how deeply telepathic, aware and savvy these creatures are. The professional dolphin-tour proprietor forbade Steve to drop orgonite into the water from their boat, so he hired a boat and did it on his own, which is when they showed up. A lot of environmentalists are misguided.

It may be the Cosgrove Shoals dolphin pod, which showed up at Sandy Key, right after Carol tied to a mooring buoy at Sandy Key, and they obviously wanted some up-close interaction, because they were swimming close by, all around, and under the boat, in that crystal-clear water.

I don't know if you've ever seen a dolphin's dorsal fin break the surface nearby, but it's a little daunting the first time. On Carol's excursion to Sandy Key that day, her son, Nick, was the first to spot them and he yelled, 'Shark!' which fairly terrified Jenny, his younger sister. Carol was reluctant to get in the water, because she didn't want to cause her kids more anxiety, but they tossed some balls, at least, and got a healthy dose of dolphin love, for a half hour or so. As soon as another boat came into view, the dolphins vanished. I had stayed home to start writing this book.

If you're energy sensitive, you can find underwater death transmitters by scanning the horizon for eruptions of DOR, from the surface of the sea up to the sky, at around low cloud level. From a distance, the area just looks smoggy to anyone else. If, like me, you're not particularly energy sensitive, you might have to trust that the dolphins will eventually take care of those, if you toss enough orgonite out. Just like with The Operators, though, it's probably not productive to second-guess their agenda. The towers closer to shore might be causing you some angst, but the massive death transmitter arrays in the deeper water, farther away, are much more threatening to the dolphins and whales, we believe.

When you get to the 'eruption' site, you'll likely see a tower, an object or at least a dense concentration of fish, close to the seabed, on your sonar scope. Fish, as I mentioned, don't seem to discriminate between DOR and POR and they generally congregate around underwater structures, anyway. When DOR prevails in an area, though, very few fish are found swimming in the open water.

Dolphins show up on ordinary sonar scopes, ('fish-finders'), as very large fish, because sonar bounces off of air chambers in sea creatures, and dolphins have lungs, so the chambers are much bigger. Affordable fish-finders show fish and objects as symbols, not photographically.

If you're not energy sensitive, just toss a towerbuster every half mile or so, throughout the smoggy area in a grid fashion. Use your GPS to track distance and direction, if you're like me. You'll probably see the smog disappear, before you even head back to port.

Sharks don't have air chambers, so they don't show up on fish finders. Sharks go where the fish are, like robbers go to banks, but they won't be seen where there are dolphins.

Jeff had been taking orgonite out to sea on his surfboard, before we moved back to Florida, and once a five-foot long spinner shark leapt out of the water directly in front of his board, as he was paddling out. I would have been daunted, but he kept going.

We had assumed that these underwater transmitters, northeast of Jupiter and St Lucie Inlets were set up to discourage dolphins from approaching shore, especially since we saw a Navy, (may have been Chinese, according to what Carol was seeing psychically at the time), destroyer, idling in those waters about five miles offshore, the first time we gifted the ocean in September.

Lately, Jeff saw a ship idling in those waters during the night, every night, but we later figured out that those transmitters seem to relate more to an occult activity, affiliated with the nuclear power plant on Hutchinson Island, 30 miles north of us. We've been doing more work in those waters and gathering more intel. Whatever was done from the ship has apparently failed to take effect, because there's still no persistent smog out there. The towers are all on a shelf that's around a hundred–sixty feet deep, and within easy range of divers.

A week after Jeff tracked that ship's nocturnal activity, he and I took a load of towerbusters to drop parallel to the beach, along Hutchinson Island, which is twenty miles long, and to particularly gift in the vicinity of that nuke. A 'diffuser' is shown on charts, extending on the seabed for two miles out from the beach, east of the nuke, which is quite close to the beach. The south end of the island is St Lucie Inlet; the north end is Ft Pierce Inlet. The nuke is approximately in the middle.

Indian River is what the body of water behind the island is called, though it's technically a long, wide lagoon, not a river. It's part of the Intracoastal Waterway, which is a connected system of channels, canals, estuaries, rivers and inlets that starts in Canada and goes all the way along the East Coast, around the entire Florida peninsula, and along the Gulf Coast all the way to Mexico. It's been a handy gifting asset for when the seas are rough, because we can arrange for the seas to be behind us, during the sea-gifting legs, and return to the boat launch facilities on the relatively flat waterways. The Sylphs seem to like it, when we gift the waterway, because some of the brightest ones we've seen have showed up during the inland legs of our sorties. Manatees are seen in greater numbers on the waterway, soon after we gift, too.

There were no fish at all on our scope along the way, north of St Lucie Inlet, and the water was quite opaque; a sickly chartreuse. When we got to the diffuser, a couple of unhappy-looking Feds were in a boat, anchored close to the beach in the rough water. They weren't even pretending to fish. Their boat was bobbing in the roiling, brown water that was coming up from the diffuser, and drifting north with the current. That water looked just like the Mississippi did, before people gifted it—opaque, bilious brown. The sea bottom in that area is white sand. I bet those two guys got pretty sick, because even though Jeff and I still had a lot of orgonite in the Zodiac, we still got a dose of radiation sickness from being there just a few minutes. We dropped several towerbusters along the diffuser's length, then a big Holy Handgrenade at the end of it, which is two miles from the beach. I was wearing my Harmonic Protector, which may be why I didn't get as sick as Jeff did, from that. Jeff wears his on gifting sorties now.

> An 'HHg' (Holy Handgrenade—I named it after the device seen in "MONTY PYTHON'S HOLY GRAIL") is a cone or pyramid-shaped orgonite device, four or five times bigger than a three-ounce towerbuster, which has a fancier crystal in it and a simple, cone-spiral coil. The shape, mass, crystal and coil get you more bang for the orgonite buck, but they're a little more difficult to make, so we get most of the jobs done with simple towerbusters, which can't be beat for creating synergistic effects, through distribution, after all.

On our way back to the boat ramp on the Indian River, from Ft Pierce Inlet that day, we saw a persistent smog bank that had just formed at cloud level southeast of the diffuser, in the vicinity of the new underwater towers, and there was a sometimes-flashing UFO weaving in and out of that suspended, dense yellowish-brown energy field, appearing and disappearing. Carol told us that this visible energy field was a thought-form, that was generated by people underground at the nuke plant, using the deadly orgone radiation produced by the reactor itself. Before long, that smog bank disappeared. I think the UFO was being used as a last-ditch effort to keep it alive. We've often seen flying saucers attempting to save deadly energy fields, and sometimes they 'refuel', at the larger death transmitters and along high-tension lines. This is pretty common, actually, so keep your eyes open, if you want to get your own evidence.

We saw a similar, but less dense, moving-against-the-wind smog bank develop, months earlier, in the same spot right, after we gifted the backside of that nuke, in the Indian River. That one was moving like an enormous flock of birds, which may indicate that it's a ritually-generated thought-form, (bear with me), that's sustained by the DOR from the nuke plant. Remember that the bad guys use DOR to get their destructive, parasitic work done, similar to the way that we use life force, (POR), for our healing work. One application of thought-forms is to influence the minds of entire populations, which might be why we sometimes feel relief, when we leave the US. Carol mentioned that the thought-form being fueled by that nuke plant, may have been used to prevent the people in the area from waking up to the reality of the damaging effects of nuclear radiation. When people aren't dumbed down, they're more likely to take notice of skyrocketing cancer rates in their area, for instance. Ordinary commercial nuke plants cause cancer, in endemic proportions, for about forty miles around, as Dr Reich discovered in the early 1950s.

I once saw one of those enormous forms moving up into the sky, from an Air Force base, north of Sacramento, California on one of our gifting expeditions, and when I asked Carol what it was, she said, 'I'm amazed that you saw it, too!' She told me it was a nuke-sustained thought-form. The reason I could see it, was probably the same reason anyone sees smog: positive-charge static field that causes toxic, (dark; solid), matter to remain in colloidal suspension, within the energy field. Orgonite in the vicinity generally ionizes these energy fields, causing toxic colloids to drop quickly to the ground, and neutralizing much of that, in the process. Orgonite cloudbusters do this in the upper atmosphere for many miles around, and you can point a cloudbuster at any smog bank, if you want to see it disappear in a few minutes.

My lovely second wife, 'X-2,' had worked at Seabrook Nuke Plant for fifteen years, before I met her in 1996, and from our conversations I gathered that very, very few people who work in nuke plants reach retirement, because they most often die of cancer, long before that happy, lucrative eventuality ever comes around. It's not a problem to shield from the effects of nuclear radiation, but it's impossible to shield from deadly orgone radiation, which extends, relatively unabated, for forty miles from an average nuke, and is most severe closer to the source, (the reactor).

I think that a nuke plant is the ultimate, parasitic creation of the occult/corporate world order, because it kills lots and lots of people from a distance silently, gradually and undetected. They found a way to do it, within neighborhoods, more recently with the new death towers, which mainly broadcast deadly orgone radiation. Think about how evenly these nuke plants are distributed across population centers, especially in Europe, and note the sky-high, cancer death rate, in those areas closer to the nukes. Now, factor in the countless thousands of smaller, unregistered, nuke generators underground, which power the new death tower and HAARP networks. The bad guys clearly intended to cause most of us to have untimely cancer deaths.

It's the DOR that sickens and kills people in the area, of course, not nuclear radiation. Nukes just happen to be the best way to generate DOR, which is perhaps why there are so many of them now. Your power bill didn't go down, by the way, in spite of the Nuke Cabal publicist prostitutes' enthusiastic claims that your power bill was going to be dramatically reduced—remember?

Your power bill also didn't go up, when millions of death towers, suddenly sprang up all around the world, in the fall of 2001. That's because the new, unregistered, underground nukes are powering up all of that new weaponry. They're all on their own underground, power grid.

Carol and I found out, in the Namib Desert east of Swakopmund, Namibia, that distributing some orgonite around a nuke-waste dumpsite immediately gets rid of the persistent smog over the site, which might mean that it's also transmuting the waste itself into harmless material. A full life of ten minutes is preferable to a half-life of ten thousand years, I think. The European nuke cabal dumps a lot of its waste into sparsely populated places around the world, like the Namib Desert. They probably just dump most of it at sea, when nobody's looking, which is yet another reason for us to give orgonite to dolphins. These are some of the corporate folks, who fund the environmental movement, which is why schizophrenic poseurs, who want humanity, (all except them, of course), to die, so that 'the beleaguered goddess' may thrive, won't ever mention it.

Inexpensive free-energy devices, made in every city, town and village on the planet, will soon do to the poisonous nuke industry, what they will also do to their Siamese twin, the Petroleum Cabal: why would anyone need to buy electricity or burn gasoline, under the circumstances? These mostly-simple inventions are already made and proven, and many have been around for over a century, contrary to what Colonel/Agent Bearden has been claiming in his charming, but enervating, narcotic techno-babble obfuscations. Cheap zappers are going to do the same thing to the Pharmaceutical Cabal, the hospital gulag archipelago and the bloated, rotting Western cadre of wealthy serial killers, (MDs).

> Obfuscate: to make unclear; bewilder—from the Latin root that means 'darken.'

As our discernment gets sharper, we begin to honor the

instinctive feeling of revulsion that comes up, when we encounter disinformants, even the charismatic ones. If we all had the discernment of dogs and little children, we'd be ahead of the game, by now.

> According to French historian, Jean Markale, the Middle French root of the word, 'obstacle' means 'devil.'

The disinformation cadre on the internet, who make up the What To Think Network's rear guard, 'strategic retreat,' campaign, have been a formidable obstacle to discernment for struggling sleepyheads, but they're ultimately just a challenge or test, for the more persistent among us.

How are we going to take down the Gold and Diamonds Cabal? We could decide to trade with seashells, which certainly have more real value than dollars do, so maybe it's a non-issue. When the French left Haiti they took all their gold, so the Haitians agreed to use little dried gourds from a rare tree and that worked as well as gold. They still call their currency, 'gourds.'

That behemoth on Hutchinson Island was the first nuke plant that Carol and I ever gifted. That was in November 2000, and we found it in an interesting way. We had just arrived in Florida from Pestilence, Texas—oops, I mean, Port Aransas, a place where the Gulf of Mexico was beleaguered by constant red tide. Why we didn't think to toss orgonite in the Texan Gulf, in those days, escapes me at the moment, especially since we expressly gifted a major vortex and Atlantean relic, on the coast to the north of Port Aransas. As I said, 'common sense' is an oxymoron. We intend to get over to the Florida Gulf Coast pretty soon, and end the perpetual red tide near Sarasota.

I was lying on the beach at Ft Pierce Inlet State Park, getting

some sun, and had the Terminator zapper on my belly. Carol saw a huge amount of blue orgone streaming up, out of the Terminator, which indicated to her that something not far away, was generating a massive amount of DOR. Looking around, she could see that the source of the DOR was to the south. As I mentioned, part of the function of orgonite is to transmute ambient DOR into healthy, positive orgone radiation. When a source of DOR is nearby, orgonite puts out more POR. This principle is the essence of how we're undermining tyranny, by neutralizing its poisonous infrastructure.

The next day I made a small orgonite device, at our RV campsite in Ft Pierce. We drove south ten miles from that beach, and there was the nuke, right beside the highway. Carol saw brown energy extending to the horizon in all directions, and she felt nauseous, but when I put the orgonite device in the bushes, as close to the nuke as I could safely manage, the brown energy field shrunk to a sphere, whose outer limit was at the orgonite device. She felt fine again, too. The sky instantly became brighter and small, white cumuli began to form, all around us.

That was our very first effort to counteract the effects of deadly technology. Before that, we had only put orgonite in a few vortices to heal them and to get their energy spinning in the right direction. That was a month before we invented the orgonite cloudbuster.

We got our first intimation of that curious transmutation principle several months earlier, as we were driving past the nuke plant next to I-95, north of Portland, Oregon. Carol was driving her car behind the Zapporium at the time, which had a lot of orgonite in it. As we approached the nuke from many miles to the north, a huge fountain of energy erupted from the back of the Zapporium, and it increased until we passed the nuke plant, then decreased gradually for many miles

past it. A year later we tossed a bunch of orgonite in the water, next to that plant, and it was de-commissioned soon afterward. Ooops. I don't think I'll be arrested for printing this, because we've disabled scores of nukes since then, (mostly unregistered ones), and told about it with impunity. Lots of folks have been doing this, in fact, and so can you.

During the following May, as we were driving through an area of Eastern Oklahoma, where tornadoes were being generated by HAARP arrays, Carol was driving her car behind the truck and frantically signaled me to pull over. She said that the cloudbuster in the back was throwing out so much orgone that it was obstructing her view. I moved the CB so that it was pointing out the front windshield of the truck and then she could see the road again. We plowed a tremendous blue furrow through the dark, HAARP muck in the sky, that day. That was six months before the death towers sprang up, all over the planet and changed the equation a bit.

The orgonite device we left in the bushes, at the Hutchinson Island nuke plant, was likely found and removed, because in those days we weren't as aware of the constant surveillance we were no doubt already under. Gifters, these days, know how to block surveillance tech and discourage pavement artists, (the 'spycraft' term for professional spies who follow other people discretely). Jeff gifted that nuke again a couple of years ago, but whatever it is that people are doing underground, was apparently neutralized, on our last gifting sortie there.

Our very first environmental 'gifting' effort was on my big brother's and his wife's, (Jim and Melody Croft), heavily-wooded property in Northern Idaho, in the summer of 2000. Melody, who is psychic, felt troubled every time she passed by a particular spot, along a trail on their land, and Carol said that an elemental was discontented about something that had transpired there, and wanted redress from humans, who had caused the imbalance, before my brother bought that particular piece of property. We tossed a few ounces of orgonite in the bushes, and after that, Melody said it felt good to be near that spot. Carol talked to the elemental and told us that he was ecstatic and grateful for the perpetual energy gift.

We apparently chased some parasitic aliens from a nearby, pinched vortex after that, restoring the vitality, form and proper spin to the vortex. It was the first time I clearly saw an alien, and he wasn't pleased. Pirating earth energy isn't only done by human parasites, of course.

In January 2002, another level of environmental gifting produced the 'flying dolphin' confirmation, when Rick Moors put the first orgonite device into the Pacific Ocean from a jetty, at Redondo Beach in LA. Sea gifting became Carol's and my main gifting interest, after she had the extended interaction with the vast pod of rough-tooth dolphins off Costa Rica, in December 2004.

Right after Carol's experience in Costa Rica, a series of confirmations started being reported from all over, including more than a few first hand accounts, of human-dolphin interactions involving orgonite.

For example, Eric Carlson, who lives in Boston and was vacationing/gifting extensively in South Florida shortly after that, tossed two orgonite towerbusters out into the surf, during a walk along Miami's South Beach. Within a few minutes, a manatee was seen where he had thrown the first one, and a blue whale nearly beached himself, a little later, in the spot where Eric tossed the second one. The whale's historic visit was filmed, and it made the six o'clock news, though nobody knew what Eric had done, and he was as surprised as anyone, by what happened.

One of our challenges, as gifters, is to recognize the genuine miracles that happen as a result of our efforts. I think we're conditioned to believe that miracles are only real, if there's a symphony orchestra, or angelic choir in the background, like in the movies.

A few weeks later, two beluga whales were seen swimming around in the Delaware River, sixty miles upstream. That was widely covered by the media, too, and those small whales interacted with a lot of people, who went out in boats to see them. A gifter in Pennsylvania contacted us to say that he'd left a lot of orgonite in that river, just upstream from there, where the water became too shallow for the whales to swim.

The beluga whale, which was seen in the Thames River in London, by British Secret Police Headquarters, wasn't as fortunate, though his companion apparently escaped being captured and murdered.

In May 2005, I spent the day riding on the ferries in Puget Sound, tossing specially made orca gifts over the rail, from Orcas Island to Seattle. Orcas hadn't been seen in recent years, and I wanted to see if they'd come back for orgonite, during the following summer. A pod of orcas showed up in Puget Sound, 30 miles south of Seattle, near Tacoma, after a fellow who lives and gifts in the Seattle area tossed two dolphin balls in that spot. The appearance of that pod made the news, too.

My first cetacean encounter was with a beluga whale. He was in a tank, at the Vancouver Aquarium, where I took my family in 1988, after we moved from Saco, Maine to Mt Vernon, Washington, an hour north of Seattle. This circular tank's top was about three feet off the ground, and you could walk around it.

As I was passing by, the little white whale raised his head out of the water and looked directly at me. I felt compelled to approach him, and when I got close, he spit a stream of water in my face and then quickly turned with a flamboyant splash, which also got me wet, and swam away on the surface. I guess that was a sort of baptism. That was a period in my life when I was just starting to break free of the occult/corporate treadmill—I hadn't yet started my own business and I was still being severely abused at home by my spouse; yes, I was a typical, programmed sap.

My next two cetacean encounters happened about a week apart in October 1995. I had just lost my family to a horrific divorce court, after being maltreated by a profoundly unhappy, psychotic, predatory woman for 22 years, then publicly cuckolded and divorced by her. Wham, wham... wham!

That turned me into a grieving wanderer, barely sane, but as I said, when The Operators can't lead us to our destiny, they might drive us. Even in the worst of my torment, I knew it was all preparation for something that I'd eventually be doing. I didn't have a clue what that could be, until years later. Sometimes we just get strong intuitive hints, when we're passing through that valley of torment, and I think those come from The Operators, as a token of their mercy and encouragement.

Initiations take a lot of forms in this life and if you're lucky yours aren't that traumatic. When we sharpen our instincts we're more likely to be led to new awareness, rather than driven. I'm not complaining; just stating some things for the record, so maybe you can get your self-empowerment work done the easier way. The following tale can be an object lesson for you in this case, grid willing.

I've never been happier or more productive than I am right now, and my future looks bright, but if X-1 hadn't slept around and then divorced me, I'd no doubt still be with her and couldn't have begun the work that Carol and I are doing now. X-1 constantly sabotaged my efforts to develop myself externally, and I chose to just take it, rather than stand up for myself—well, until the end. The day after I gently took a stand, she told me that she would divorce me, which she did, after another year or so of torture. If you're married to an incorrigible saboteur, you have my deepest sympathy, but if you won't stand up to that one, in an appropriate way, you're not going to move ahead, either. Why wait until you die to earn your freedom?

I simply assumed, in my ignorance, that the purpose of life, in this transient world, is to go through whatever torment, imposition and strife is meted out, with as much contentment, faith and detachment as one can develop, in our short span of years. I later learned that this is the same Luciferic, dualistic, mental programming, which enabled the Church of Rome to create the dark ages in Europe, so that personal attitude is anything but innocuous. The way I was, I'd have been one of those bystanders, who didn't lift a finger to oppose the tyrants, who rounded up innocents in twentieth-century Germany, Russia, China and Cambodia and herded them to the killing fields. I'm getting a whole lot of satisfaction, now, from effectively standing in the way of their American counterparts, who dearly want to do the same thing here now.

I've always known that our best personal, spiritual achievements are usually the result of having successfully acquired genuine detachment, which is usually gained through trauma. I loved X-1, in spite of her constant abuse, and I wanted to keep our four kids from getting the worst of it, so I allowed her to direct it at me. I've never, before or since, encountered anyone with

such deep and sustained hatred, as this gal had displayed toward everyone who was close to her.
The fact that a life of personal sacrifice, as epitomized by Father Damian in the Molokai leper colony, and also by parents of severely handicapped children, who opt to care for them personally, is an honorable one, muddied the waters for me spiritually. In those cases there was simply no other option, except to expend one's own life energy, in order to sustain others. I would have stayed with X-1, of course, if I hadn't been cuckolded and cast out, the latter being done under the auspices of the Federal Reserve Corporation's court and leg-breakers.

I was a fool, for not taking the children from her, by reporting her physical abuse toward them, and there's no getting around that. I won't make excuses for myself. What I knew was at risk, if I took that course, was that the Child Protective Services agency in our county, in cahoots with all of the judges, were in the business of kidnapping blond, blue-eyed children, when any report, (real or false), of abuse reached them. Our kids all had those physical features, so they'd have gotten top dollar, on the open market. I thought I was in a no-win situation but, with hindsight, I really ought to have taken the abusive mother to court, during the early part of the divorce process, and done the deed for the sake of my children, regardless of the threat.

In an old, Persian poem, by Rumi, there are two young lovers, named Majnun and Layli. Majnun has lost Layli and seeks her everywhere. He goes through untold suffering and danger and, pursued by watchmen, (cops), he scales a garden wall, throws himself down on the other side, exhausted, and finds Layli. It's a good parable, which describes the process that one often goes through, to find a new truth or level of awareness. This illustrates how one properly feels about seeking truth; it's a drive as essential to humanity as the drive to find a mate, in fact—more essential, really, since we can all be happy without physical love, but nobody can be dynamically happy in complete ignorance.

Thanks to the 'hundredth monkey' paradigm, which I believe is valid, when enough members of a specie get to the comprehension of a new truth, everyone else automatically gets it, too, more or less. The new truth that my own suffering eventually led me to, is that 'the meek are inheriting the earth,' and it's likely that you won't have to slog through the pure, personal hell I had to go through along the way, to this realization. You're welcome!

I had completed a seaworthy little boat, during the year after the separation, prior to the divorce, when I was living in my funky little sign shop on the outskirts of town. After I failed to sell the successful business, (I couldn't bear to see her with her boyfriends, because I still loved her, against all reason), it seemed logical, in my torment, to eventually just sell off the equipment, and take my boat on a trailer to warmer seas—a familiar, remembered comfort during my youth in the West Pacific Ocean, that looked pretty appealing in my debilitating torment. Being on the water tends to heal us.

The reason I couldn't sell the business, is that almost nobody was able to do hand lettering any more, and my reputation was built on my creative talent and my hand skills, not on the computer-generated vinyl signs that had come to dominate the sign industry, during the previous decade. I did that work, too, of course, because 'work no due; no work due.'

If I'd been able to sell the business, I could have set up again in an area far enough away, that I could see my kids, without having to run into X-1 and her tag-along beaus, every day. It was a pretty small town, after all, and I just couldn't seem to get detached from my torment. Much later, after I stopped desiring X-1, I realized that it wasn't the boyfriend thing that was stealing my energy; it was simply being within her convenient reach that bled me out, etherically. That never happens accidentally; it's intentional. There are some people in this world who thrive on stolen energy. I bet you know one or two of those, because they usually assume positions of authority or influence, within the receding social and institutional paradigm. X-1 eventually married a multi-millionaire, for instance.

Some guys can tough it out and re-create their lives under those circumstances, but I simply lacked the strength or skill to do that, frankly. Every contact with her, even on the phone, left me as vitally flat as though I'd been assaulted with a cattle prod.

D Bradley is going through that, right now. Even though he, too, was cuckolded and cast out of his own nest, where he had been a stay-at-home, loving and watchful dad to his four little sons for six years, he stays nearby, living in his pickup and giving all of his money to his Ex, who has the support of her well-to-do parents and, presumably, her lover, who is allegedly a successful architect. I know he loved her as much as I loved X-1, but maybe the phenomenal personal discipline and training he acquired in his youth, from Torkum Sassarian, one of the dark masters of the Great White Brotherhood, is actually an asset right now. In that case, the training was designed to make him a good research tool for the dark masters.

The Feds made sure he could never get employment, have a passport, or re-establish a capital-based business, after he turned against the Great White Brotherhood, and was later shot and killed by a government-employed sniper at close range, in 1999.

On top of that, the occult/corporate machine are doing their best to erase him from public awareness, again, so I hope you'll purchase some of his peerless orgonite wares from his website, www.thevine.net/cbswork. Also download and watch his free, half-hour documentary film, "CHEMTRAILS: CLOUDS OF DEATH", from www.ethericwarriors.com or www.worldwithoutparasites.com. He never intended to make a nickel on that film, so it will always be given away.

I was rather consigned by circumstances and my own weakness and instability, to live in my car and travel around, painting signs for subsistence, too depressed to stay in one place for more than a week, and too poor, at any rate, to have a home. During that time I encountered an enormous, hidden segment of society that I'd been unaware of: 'deadbeat dads;' fellow wanderers—the new hobo class. This is the sort of thing that can happen, when we don't stand up and do what our conscience dictates, which in my case would have been to fight for the custody of my abused kids. Bradley had no choice in his case, because the deck was entirely stacked against him. His ex is a good mom, so that's a small comfort, at least.

Once you start to run up an alleged tab with one or another National-Socialist bureaucracy, it's a one way ticket to penury, in most cases, because these non-crimes were federalized, shortly after the Bureau of Alcohol, Tobacco and Firearms blew up the Murrah Federal Building in Oklahoma City. When the tab gets to a moderate level, a fake warrant is issued for your arrest, in that state. I guess that, at least, hasn't been federalized, yet, but I think it will, be before long. I know the warrant is fake, because when those predatory cops in Las Vegas molested me on behalf of the CIA, five years later, no warrants showed up for me, on their closed-circuit internet crime computer, and they had to release me, much to their chagrin. The exotic, very quiet helicopter that was hovering nearby, facing us throughout the episode, banked steeply and flew away pretty fast, when those two thugs got back in the patrol car, so somebody else apparently wasn't satisfied, either. The sewer rat agencies pretty much stopped painting their omnipresent helicopters black, by the way. I think, a lot of them were getting shot down by patriots, before, and an awful lot of people were starting to notice them.

That reminds me of something a Persian refugee of the London-sponsored, Khomeini regime told me, in the mid-1980s. Apparently, the Persian Gestapo equivalent, who were charged with rounding up countless thousands of innocents, for firing squads and one-way prison trips, eventually stopped wearing their terrifying uniforms, because too many snipers were shooting these thugs in the street. The Baha'i families in Iran, whose members are shot by firing squads get billed by the government for the bullets.

When National-Socialism was being introduced in America, during a managed, global economic collapse, in the nineteen-thirties, the first wave of hoboes was generated. These were the fathers who could no longer get employment, and wandered the land looking, (mostly without success), for work. In those days, the corporate court infrastructure wasn't yet set up to start billing these wanderers for 'child support.' Even if that had been set up, I doubt people would have wanted it, because marriage was still considered sacred, then, and divorce was only granted on the basis of genuine, spousal abuse. Marriage was also not generally considered to be a government matter, either, and rightly so.

Big Brother's new 'charity', suddenly put the state in the role of provider, then—a body blow to the nuclear family, from which Western culture hasn't yet recovered, in National-Socialist countries. It was during that period that the Federal Reserve Corporation insinuated itself into marriage and family life.

It generally takes ten years to get back on your feet, after a personal economic calamity, such as the loss of a home and business, through divorce or uninsured fire. If a physical calamity happens through an Act of God, you can get a low interest loan from Big Brother, but a discarded father has no recourse, after dispossession, unless his parents are charitable—not often the case in the stronghold of materialist America, sad to tell.

On top of that devastation, when a government bureaucracy, set onto a disenfranchised father by a corporate judge, decrees that he owes them a lot of money, on behalf of his own kids, they make it impossible for him to have a bank account, driver license, employment, (on the books, at least), passport, etcetera. What's hardest to bear is that his kids get the worst end of it, because they're deprived of a father's physical as well as financial support, due to arrest warrants, and these unfortunate kids often have to endure a string of Mom's live-in, essential male prostitutes, some of whom presume to play the father role. I'm sure you're aware of this common spectacle. Sometimes the gals find decent spouses, eventually, but really, we sober men don't generally want to be economically and emotionally saddled by capricious, gay divorcees and their unhappy children, because we know it's a form of castration. Some things just can't be sold with sex.

Pajama People are well-conditioned, by the What To Think Network, to believe that all of this is the father's fault. Single moms in the West, are represented by the What To Think Network as super-heroines, 'victimized by heartless men no more!' and this mindset makes it even more unlikely for them to find decent mates, after their capricious divorce.

The blind acceptance of this paradigm, by the PJ folks, parallels their blind acceptance of the largest prison system in the history of our species, and its blatant institutional racism.

Both of those developments are genuine calamities, and the heartbreaking part is that they're not 'acts of God;' we all allowed them to happen, through our own neglect.

When the Gestapo were rounding up undesirables, (including Jews), in National-Socialist Germany, most people, even outside Germany, chose not to notice, because they were beginning to enjoy relative prosperity, finally. Adolf Hitler was on the cover of Time Magazine, in those days, as Man of the Year and so was Benito Mussolini. This National-Socialist American regime routinely rounds up men of color, who show leadership potential and disappears them into the Gulag Archipelago, and the What To Think Network ignores it, so white PJ folks, (the majority), just ignore it, too, and they automatically assume that 'black, Indians and Hispanics are simply more prone to being criminals; that's why our prison system is so big'. European Pajama People, under Nazi rule, mostly assumed that Jews, Gypsies, homosexuals and [alleged] commies, all had it coming, too. Most of the people, at least in Germany and Austria, who weren't in those little demographics, didn't feel threatened at all, by the Gestapo.

Do you see how this works?

There are a lot of parallels between National-Socialist Germany, in the 1930s, and National-Socialist America, since World War II, especially lately, unfortunately. The 'useless eaters' in America, are mostly people of color, and the often-unwitting Europoid committers of 'crimes against the state.' In case you don't know, that's a newspeak euphemism for 'victimless crimes,' or 'non-crimes.'

X-1, who hadn't worked in twenty years, started her own business, right after our separation, with the money I'd been giving her, (most of what I earned—until I just couldn't do it any more), and she was already doing as well as I had done, by the time I went into exile, a year later. Selling our home, which had acquired a lot of equity, gave her and the kids even more material security.

The zapper ended my depression, but not my grief. At least I was able to start creating a stable life for myself again, after I destroyed the parasites in my brain, one day in April 1996. Within another couple of years, I had a tiny apartment in a neighboring state, and my two younger kids were sent to live with me, apparently so X-1 could have a little honeymoon, in a subsequent marriage. She and her new hubby had the Sheriff Department take them out of my home, after a month or so, which pretty well tore me up again.

Not being depressed any more, and since I was five hundred miles away from X-1, I at least didn't miss a beat, economically and the grief of losing my kids again, didn't debilitate me, that time. I continued to grow my little zapper business on the internet. By June 2000, Carol and I got together and developed the Terminator zapper, which is said by many to be the finest zapper on the market. It's the best-selling one, though we've never advertised or promoted our zappers in conventional ways.

With the relatively quick, moderate success of the Terminator the zapper, business started to take off enough to modestly support Carol and I, though we couldn't yet afford to get our own place. Before I showed up in Idaho in May 2001, she had decided to leave her abusive ex-husband, whom she had divorced years earlier, but had agreed to re-join under common law conditions, for the sake of their daughter, Jenny, who was fourteen when Carol and I got together. Jenny opted to stay with her dad, then, and joined us a year later, when we finally got a place. Carol had induced him to stop drinking, when Jenny was a baby, as I mentioned before, so at least she was physically safe there.

Part of why we lived on the road for the following year, was that the government had started going after zapper makers pretty aggressively, and we felt safer not having a physical address, under the circumstances. As soon as that trouble passed, we settled down, and by then we could afford to keep a home, thanks to our business having become more stable and predictable. Jenny underwent some trauma during that time, living with her dad, that wouldn't have happened, if her mom had been around, and Carol and I both feel remorse about that, nor do we feel inclined to excuse ourselves for letting it happen. The trouble mostly came from Jenny's choice of friends, not from her home situation. She felt that her mom had abandoned her, and that no doubt influenced the choices. Jenny seems to have worked that out by now.

I don't know if you're aware that it's extremely risky to be a teen, these days. The risk is imprisonment for a wide variety of non-crimes, which police are induced to enforce with rabid vengeance. This is all part of the second calamity, the American Gulag, but look at how intimately it's joined to the first one, which is the destruction of families by the occult/corporate courts and the What To Think Network.

I heard about a guy, who started a sizeable trust fund for his kid when he was born, which wasn't for college tuition; it was for eventual therapy. That's one of those little stories, which are funny, but not funny. If you carefully look at most family troubles, these days, you're likely to see the hand of the occult/corporate hierarchy somewhere in the mix, steadily stirring.

The fact that Carol and I are both on marriage number three, can be attributed to neither of us having been trained in early childhood, to think critically, or to learn how to distinguish genuine friends from parasites. Every child should learn these lessons, but if you, like me, were fifty-one before you finally got it, I say, 'Good for you!' Fortunately for us both, our commitment to this marriage goes beyond just loving each other, and we love each other an awful lot.

Carol's X-2's abuse toward her and others was verbal, emotional and psychological, never physical. Most folks assume that physical abuse is much worse, but in terms of child development, maybe it's less damaging than the more subtle forms of abuse. A spouse always has the option to leave in order to stop abuse by someone who's incorrigible, but children can't make that choice.

I was often at the receiving end of gratuitous, physical violence from X-1, as our children also were, when I wasn't around to absorb her rage. I think that if the government hadn't become aware of X-1 sticking a fork in our baby's head, after the divorce, I might have had gotten worse treatment from that pedophile, kidnapping, judge later on, when I was finally subpoenaed for not paying the state money for my kids. I never reported her abuse of the children, when we were together, because I didn't want them to be kidnapped by Child Protective Services, as I mentioned.

I personally knew several people in the area, who had lost their children to this satanic agency and the county court, so I was properly terrified into being a sap. Washington State is notorious for institutional kidnapping, especially in the county we lived in. I simply did my best to shield them from her, during her almost-daily destructive rampages. My biggest source of grief, after the separation, was not being able to shield them any more, especially because she'd become an alcoholic by then. The sense of powerlessness drove me close to suicide, but I knew that suicide emotionally scars one's children even worse than divorce, or an alcoholic parent does, so I refrained from doing that.

My ego was obliterated, in the process of all that trauma, and I needed to rebuild it, in order to have a productive life again. Reconstruction of my ego was a happy process, facilitated by my friend, James Hughes, and I'll describe that in another chapter.

Carol and I spent our first year together, on the road in the camper, which we started calling the Zapporium, since it was also our factory.

It was prudent to be on the road at the time, because the US Government was going after all of the zapper makers. They had just put our US competitors out of business, through open intimidation and extortion. They even threatened one of our zapper distributors, who went to ground, as though his head were about to be pinched off by Big Brother. I think that if that fellow hadn't been a marijuana addict, he would have responded in a more balanced way. Those federal traitors never even emailed us, and by then I'd been publicly advocating, from my email podium, the destruction of this unlawful, National-Socialist, corporate regime. I was corresponding with hundreds of folks, many of whom were critical thinkers.

I guess the US Government prefers non-moving targets, though they've more recently tried several times to kill Carol and I, on account of our later international networking and environmental healing successes.

Until the gifting network developed some momentum in America, the federal agencies were winning the generations-old War on Healers here, but by now their human resources are spread so thin, that they can't seem to develop any effective predatory agendas lately. More on that later, concerning our good friend, the eminent natural physician and research pioneer, Dr William von Peters and his continuing, prevailing survival, against the odds.

Lots of people in the US are making and selling zappers now, I'm happy to report. Many of them are copying ours, which is quite flattering. We keep a few of the subtle energy components of our Terminator proprietary, of course, so that we can keep an honest edge in the market. Ever since Clinton, at Bush Senior's behest, gave Beijing the US Patent Office, it's been pointless to try to patent anything, and I've long felt that getting an endorsement, or any sort of 'help' from this treasonous regime, is as useful or desirable as entering beneath the protective shadow of the National Man Boy Love Association, or the Ku Klux Klan would be.

Back to the October 1995's dolphin encounter:

I was heading out into the Gulf of Mexico from South Padre Island, Texas, on my way to God-knew-where. It was tough getting out through Brazos Santiago Pass, because the tide only ever comes in there—never seems to go out—and the wind was against me, too. If it weren't for the little electric trolling motor, that I brought along for these situations, I'd have been stuck there; I barely squeezed out past the jetty into dark, open water, then started a long tack, to the northeast along the shore on my way east, (where the wind was coming from), across the Gulf of Mexico to Yucatan, in the shorter term and, hopefully, at least to Belize. I much admired the Black Caribs of Dangriga—my inspirations for a community's independence in a hostile world, some of whom I'd gotten acquainted with, on a previous visit. They're quite intriguing, intelligent and resourceful people, who deeply revere their tribal identity and culture.

I loved Belize, when I'd gone there for a couple of weeks, a year before, and I wished to try my hand at recreating my life there, based on my sign-painting skills. X-1 had intimated that she was interested in getting back together with me, and going to Belize, much as we'd gone to Tonga in 1984-5, when our two older kids were small. Ever since our first was born in 1979, what I most wanted from life was to get my kids to a safer country, because I felt sure America was going to turn into a bloody, oppressive National-Socialist police state, before long. When I set out across the Gulf, I was more adamant than ever about finding a safe haven outside the US for my kids, because the Feds had begun openly rattling their sabers about martial law, after they blew up the Murrah Building.

During the hours it took to tack, (sail into the wind by constantly changing directions, diagonally, to windward), upstream through the narrow pass, I was surrounded by dolphins, whom I could clearly hear, but couldn't see, because it was the middle of a moonless, cloudy night. I went out at night, because I didn't want the Coast Guard to see me. I was feeling pretty contrary in those days, and had opted not to license my nineteen-foot, modified surf dory. I was so tired after all that suspense, that I fell asleep at the tiller, shortly after I reached open water. In a little while, I was wakened, when the centerboard dragged the bottom close to the shallow, breaking surf, at a beach on Padre Island, northeast of the pass. Fortunately, the water was deep enough, to make a turn and head out.

I wasn't very appreciative of my relationship to the world in those days, and I was deeply depressed and grieving, so I didn't make the most of that dolphin encounter, though I had a distinct sense that they wanted to communicate something to me. I was deeply touched by the next encounter, about a week later. Looking back, I'm sure the dolphins in the pass did communicate something to me, on a higher than conscious level. It wouldn't surprise me if they knew, then, something about what we'd be doing now.

When the next cetacean encounter happened, I was becalmed in the middle of the Gulf of Mexico, in my little open boat. I took advantage of the inactivity by reading, and developing my tan, (all over). There wasn't a cloud in the sky, that afternoon but suddenly I was in the shade. I looked behind me and saw a large eye, also a long, wide fin raised over me. After a moment the whale, which was much bigger than my boat, silently moved forward and sounded. I felt a little stunned in a nice way and thought, 'I could have reached out and touched him!' I went back to reading, and it immediately happened again. That time, too, I just stared into the big eye, transfixed, for a few seconds until the whale sounded.

I knew something transpired between us, but it wasn't until three weeks later, when the hurricane arrived, that I started to get some meaning from the phenomenon.

By then, I realized that being out on the water and alone wasn't easing my pain, but at least I had a sense, however inaccurate, that I was getting somewhere. Vagabondage isn't freedom. Freedom is entirely internal, though it can be a lot of fun to travel when we feel balanced.

I spent a week and a half in Yucatan, repairing my broken rudder hardware, (I had to chase around Merida, looking for suitable hardware—that was fun), and getting acquainted with some nice Mayan people, before I resumed the journey. I got into the hurricane a couple days after I left and Sisal, which was the first Spanish settlement in Mexico. Merida is about an hour's ride inland from there.

After that storm passed, I started to feel grateful for my life, for the first time. I never felt threatened during the storm, strange to tell—mostly just exhausted from the constant slamming, after the boat got launched from the wave tops. I stayed dry under the boat's streamlined, roomy and removable hard cuddy, which was open in the back, and the boat steered itself very well into the wind, when I lashed the tiller to the mizzen sheet. I'm still pretty proud of that design and engineering feat. I never even had to reef the sails of the two-masted, modified Chinese lug rig. I still love little ocean boats.

The depression stayed with me until I got my first zapper, six months later. Brain parasites mess up untold millions of people these days, thanks to antibiotics' widespread and indiscriminate use. Antibiotics make it possible for parasites to take up residence, throughout the body, rather than mainly in the colon, as before. You can imagine that colonies of worms, bacteria, viruses and fungi, all excreting ammonia, formaldehyde, alcohol and a laundry list of other acidic toxins, into the brain and endocrine system, will make the host feel 'off.'

All of the parasites in the brain are destroyed, in the first twenty minutes or so of zapping. I've never been depressed, since the first zapping session, though I continued to grieve the loss of my children for years after I gradually stopped loving their mom. I still feel stung by the loss, even though they're mostly grown now, and I can finally help them all I want.

The loss of my children and home set me solidly against the Federal Reserve Corporation, whose court system I hold accountable for that nation-wide calamity. If this unlawful regime hadn't been allowed, by my grandparents' parents, to insinuate itself into our personal lives, that unhappy woman would have had to negotiate a workable settlement with me regarding the kids, rather than just set the treasonous dogs on me.

She and her clever, aggressive, man-hating lawyer were simply enthusiastically doing what millions of unhinged, unhappy, sociopathic, American women each year are programmed and triggered to do, I reckon. This sort of behavior seems similar, to me, to the way my parents' generation, the Depression Babies, gave no thought at all to allowing their boys to be rounded up by the Federal Reserve Corporation and sent to Vietnam, as cannon fodder.

Happiness is an entirely internal process. It's an achievement, as freedom is, and we each ultimately choose whether to be unhappy or happy in this life, just as we choose whether or not to be free, or to exercise discernment. A simple truth is that we can't exercise these birthrights, unless we take care of business in the mundane world. In my case, my happiness and freedom requires appropriately opposing tyranny, which I've been doing since the middle of January 2000. Before that, all I wanted to do was escape from tyranny with my children.

Anyone in the West, who won't choose to take the high road, will automatically gravitate to the easy one, because we're constantly under the thought-police pistol, wielded by the What To Think Network and other National-Socialist institutions. The adult Jews, Gypsies, Poles and other victims of National-Socialism in Europe, in the 1940s, who tried unsuccessfully to "go along to get along", and died in the ovens and work camps, can't hold a candle to the ones who chose to oppose tyranny, instead.

It's not much different now, unfortunately. If you doubt it, read some of that horrifying legislation that the US Congress is cranking out daily, and the What To Think Network's constant blather about the Homeland Security Abomination, whose eagle logo is the one the SS also uses. A former KGB head runs that treasonous agency, and his lieutenant is a former East German STASI boss. What will it take for you to recognize tyranny, short of getting rounded up at 3AM, by jackbooted, black-uniformed American cops, who always look and act like they're in a war zone?

This life is all about making choices, and nobody ultimately stays on the fence. These untold millions of disgruntled gals aren't essentially evil; just kind of intoxicated by the false hope of an easy life: plenty of non-committal sex and economic 'freedom,' underwritten by discarded husbands. They usually end up with worse lives than they had before their casual divorces, unless they're predatory and irrationally cling to the hope that they'll turn into the beloved super-moms, who they see every day, on the What To Think Network and in movies, but never in real life.

During our twenty-two years together, X-1 and I never used drugs and she only got sloshed twice, otherwise neither of us drank alcohol at all.

She immediately took to alcohol and drugs, as soon we separated, and became a fixture at several local taverns, where she met kindred spirits of both genders. After a couple years of that, she settled down and cycled through another unwitting husband. X-1 eventually married the part owner of a couple of nuclear power plants, and was coerced to sign an ironclad prenuptial agreement. They've been together for a couple of years. My kids say she hates being in the marriage, but I suspect she'll remain in it for the money. Now, she's on three prescribed brain drugs, so she can behave in a way that won't get her provisional marriage contract cancelled.

I assume you know that 'marriage contract' is an oxymoron; marriage is a covenant and no government, especially no damned corporation, has the authority to regulate that. Nor is it any of the Federal Reserve Corporation's courts' business, which is by definition strictly commercial. Their courts are business courts, not lawful ones. There haven't been any courts of law in the US, since the mid nineteen thirties, of course.

If X-1 blows this, she'll leave the partnership only with what she brought into it, plus whatever she can squirrel away from her allowance. She sold her successful business, probably as one of the conditions of the pre- nuptial, so is doubly dependent on the guy's tolerance. It would be turnabout, if he discarded her, and fair play—one for two—but I don't wish it on her, because, 1) I'm not vindictive; 2) as long as she's on those brain drugs and has plenty of cash, she's less inclined to poison my life or abuse our kids.

Two of my kids, who dread visiting their mom, on account of the step-dad, told me, "He's not human." He's an alleged fundamentalist Christian and a member of the notorious 'Millionaires Club', in a major city, so I don't imagine that he's much fun to be with, notwithstanding the 'marriage' contract. He doesn't want my kids around, which suits them and the mom. I can finally take an equal share in supporting them financially now, strange to tell.

It's just as well that I don't remember his name, because he might want to sue me for saying these things, if I mentioned it. I can't imagine what he's getting out of the marriage, because a lot of newer models are probably less expensive and more user-friendly than the cranky, older model he has now, but there's no accounting for taste, of course. Maybe he enjoys tormenting her. Controllers hate being dominated. They get especially juiced, when they can control other controllers. Maybe that's the secret to their partnership. Maybe it's none of my business.

If I don't mention these personal things, the serried rows of my personal enemies may try to take some initiative and bring all of this up in their own context, perhaps with X-1's able and willing help. That might ultimately cause quite a few people to decide that gifting is just a waste of time, on account of 'that hypocrite, Don Croft, who is promoting it.'

Fortunately, I've had some practice countering character-assassination tactics and strategies, during the formative years of the network, so I've acquired some skill at peeking around the bend, to see what new agendas they're hatching against the network. Also, my enemies are professional disinformants; they have no personal motives and that means they lack passion and charisma, so they generally have to be content with sniping, whining and finding weaknesses to exploit.

During the last months that X-1 and I were together, when she was publicly cuckolding me, she tried incessantly, (and unsuccessfully—why would I hit the woman I loved?), to persuade me to strike her, in front of the kids, (witnesses), and I only figured out why she did that, after the divorce. I'm not always a quick study, but I'm getting street-smart, these days. I think her man-eating lawyer encouraged her to do that.

Most discarded dads never abused their wives, of course. Most of us love our children, nor are we drunks or otherwise degraded; we're just grist for the impersonal, National-Socialist family-destruction mill.

The tyranny of parasites requires the continual destruction of nuclear families, in order to consolidate their political influence through successive generations, liberally reinforced with 'bread and circuses,' which is one of the What To Think Network's functions. Welfare, Medicare and Social Security are the other arm of that campaign.

I suppose soylent green is the type of bread they'd prefer to distribute, rather than welfare, Medicare and Social Security, but they've always had to move slowly, incrementally, to trick us into giving up our birthrights by degrees, rather than all at once, which would wake up even the Pajama People, and would induce them to act decisively against parasites.

Henry Ford once said,

> 'It is well enough that people of the nation do not understand our banking and monetary system, for if they did, I believe, there would be a revolution before morning.'

Political conditions are a whole lot worse, now, than when he wrote that, and the conditions in Ford's day were a lot worse than they were in America, in 1776.

If you're one of countless millions of lockstep, western democratic liberals, you probably rolled your eyes, when I mentioned Henry Ford. That's because your Marxist professors didn't tell you that Henry Ford acted aggressively, and at enormous personal expense, to prevent the outbreak of World War I, and he was the first auto manufacturer to hire black workers at the same pay as white workers. He paid them all enough, during the height of the Depression, to be able to buy the cars they were making, when nearly everyone else was too poor to afford a car. The Ford workers were the only ones who never unionized or felt a need to strike. Chevy workers were machine gunned by FDR's new federal police, when they went out on strike in those days. FDR was the premier western democratic liberal.

What did your Marxist professors do to help anyone or to make the world better? Isn't it a good time for us to look for new role models, rather than the ones presented by our professors and by the What To Think Network?

Bush Senior, who has been America's hidden dictator, on behalf of 'the new world order' since 1980, once said on the What To Think Network, 'If this were a dictatorship, it would be a heck of a lot easier!'

The demise of my family wasn't personally directed at me, of course. I simply chose to take all of that personally, which is what eventually motivated me to fight this insidious tyranny, after spending a few years seething with impotent rage, against this lawless, commercial court system and X-1's accommodating treachery.

The alternative, (the easy road, psychologically), would have been to consider myself a 'deadbeat dad,' like millions of other hopeless men, and continue to live in my car. I'd do that until I could parasite on some other poor, discarded schmuck who, 'stands up' to help pay for my upkeep, in his children's home, while I send all of my meager earnings to X-1 and her then-current john, who was sending all his earnings to an ex-spouse, ad nauseum. But I'd rather live the rest of my life without sex or companionship, than stoop to playing 'musical spouses.'

By now, can you see how surreal this National-Socialist family dynamic is? Do you see why I consider, 'common sense' and 'conventional wisdom', to be oxymorons? How bizarre, heartbreaking and tragic do you reckon future historians will find this sad, national calamity to be?

By the time January 1, 2000 came around, I still wasn't deeply committed to destroying tyranny. I simply felt that it was about to reach its zenith: an overt prison colony after roundups of dissidents, into summary execution facilities and bloated concentration camps, all policed by Russian and Chinese 'UN

Peacekeepers.' I worried that this might suddenly prevent me from being able to take care of my kids, so I moved back to be near them, and made tentative plans to get them into nearby Canada, at least, if things fell apart. By then I'd built a seaworthy boat, again, and they were just across the water from Vancouver Island. I discussed it with their mom and she wasn't opposed to it, which I knew could mean just about anything. She promised not to have me arrested—I was still stupid enough to believe that she kept promises, which might have caused her and her then-current husband some private enjoyment.

It wasn't for two more years that I discovered, with Carol's timely help, that the end of parasitic, corporate tyranny might be accomplished through the intelligent and timely distribution of orgonite, the magic bullet. A couple of years later, we inadvertently discovered that we could routinely interfere with occult/corporate mass murderers, who were in the process of planning 'terrorist' events.

A couple of weeks in a corporate jail cemented my resolve to dedicate my life to ending the rule of parasites. The humiliation of being thrown in jail for a non-crime is, what made it entirely personal for me. 'Whatever lights your fuse,' eh?

In my manipulated ignorance, I was afraid that tyranny would jump to a new, overt level at midnight on December 31, 1999, so a couple of months before that, I parked my homemade camper and twenty-four foot motor skiff by a friend's machine shop, in Anacortes, Washington, next to the waterfront. My kids were living in that town by then. I'd been living in Oregon for a couple of years before that, and I wanted to be available to them, just in case, as I mentioned. I had been promised by a pedophile, kidnapping judge in that county, that if I was ever seen in Washington, I was going to go to jail for being a 'deadbeat dad,' but I knew that the cops didn't care about my being there unless X-1 made an issue of it, to that particular black-robed criminal.

I'd moved up the economic ladder, from living in my car, to having built a lovely camper shell, onto the back of a 1972 Ford pickup, (I traded sign-work for the pickup in Nevada), and a nice, sea-kindly skiff of my own design. The latter could plane at ten mph, with a five horsepower motor and a full load, and was stable and dry in rough water.

But I still wasn't yet making enough money to have a proper home for my kids. I think X-1 and her then-current husband, (a retired Navy officer), also felt that there might be chaos on January 1, 2000, because they had me arrested the next day, when it was obvious that no computers had crashed. An hour before the cops showed up to haul me away, early that Sunday morning, those two went by the camper on roller-blades, paused to look at the license plate and as they were moving away, X-1 turned and gave me an enigmatic smile. If I were a little smarter, I'd have left right then, but being that close to X-1 for the past few weeks had made me a little punchy from lack of energy, so I wasn't feeling too sharp.

I'd been by their house to get the kids, in the previous couple of months, for daylong excursions, though more often than not, they changed their minds by the time I got to the door, and wouldn't let me take them.

I had told them where I was staying, but they apparently wanted to make sure I was still there that day, so the cops wouldn't have to hunt for me. I think X-1 relished the stupid smile on my face, as I waved to her that morning. As I said, I'm a slow study, sometimes.

I had participated in a firewalk in Seattle at midnight on Friday, December 31, 1999 and there was still soot between my toes, when I was impersonally tossed in the county slammer. Two more major, personal initiations within two days! It was getting easier, though, because I wasn't particularly depressed any more, and I was starting to have a life again.

My kids' then-current step dad, who was an okay guy—just a dedicated Pajama Man, who was engaged in playing 'musical spouses'—showed up in court with his wife and their lawyer, to try to induce me to sign away my parental rights, which I refused to do, of course. It was funny that the corporation pretended that I had parental rights at that point, but at least I didn't give it all up voluntarily—they took it from me. I'd be a real chump to formally agree to that sort of thing.

Before a year was gone, she took this john to the cleaners, after he got a large settlement for a minor auto accident, then she divorced him. She likes money a lot. That poor schmuck is probably with another gay divorcee by now, or the one after that, and still playing that stupid game, just to get regular sex. PJ folks usually drop their cues, and continue to play 'musical spouses.'

I lost most of my body fat, during my fifteen-day 'hunger strike', and felt pretty trim when I got out of there. I think it would take a month in the slammer, or on a yuppie fat farm, to take off my current, 50-pound spare tire.

I noticed that there were only a few real criminals in that jail; most of the guys in there were 'guilty' of vague non-crimes, and the strangest part is that they all assumed they were guilty, simply because they were in jail. A comedy writer can't invent anything funnier than that.

The few real criminals in there, stood out in sharp contrast to the majority. It's just like the way discarded, but responsible fathers who fail to recover their economic autonomy and become involuntary hobos and suicides, most often assume that they're 'deadbeat dads.'

Talking to those not-guilty jailbirds, most of whom were quite young and on their way to prison, was a pretty bizarre, but enlightening experience and a lesson in how pervasive and effective mind control and the What To Think Network is, among PJ folks.

That jail, like every other one throughout the Land of the Free, gets $300 per night, from the Federal Reserve Corporation, for every occupied bed, and all county judges in America get to keep thirty percent of whatever fines they're able to levy. That's a pretty good incentive program that the Fed offers these treasonous jailers and judges. There are no strings attached to all that federal money. The money is the string that ensures that there will be no local, state political or economic autonomy for the duration.

A friend of a friend who is an ex-cop, came to visit me, when I was in there, and he promised to help me turn the tables on X-1 in court on my own, without a lawyer. Visiting the jail must have been a little uncomfortable for him, because he was nearly beaten to death by Sheriff deputies in there, a couple of years before, and then they tried to throw him off of the top of a nearby tower. These unlawful courts and their gun-toting leg-breakers don't take kindly to anyone invoking the rule of law, as this brave man had done successfully, and is probably still doing.

The particular judge, who had me incarcerated, is deeply implicated in a long-standing, murderous and lucrative kidnapping ring, with the local Washington Child Protective Services staff, and a couple of psychiatrists. I knew that from before, when I met several parents in that county whose children were stolen in that courtroom. Knowing about this, in fact, made me even less inclined to give the government money, after he publicly humiliated and attempted to extort me.

I stood in front of a packed, winter courtroom that midwinter day, in an orange, short-sleeved jumpsuit, in shower shoes, with my arms manacled to a leather belt around my waist, one arm being held by a Sheriff Deputy. I ignored the cheap prostitute who called himself my 'court appointed attorney.' On my release, they presented me with a bill, $335, for his 'services,' which I threw away.

I didn't have the heart for the kind of sustained courtroom confrontation that this well-meaning ex-cop was recommending. I knew it would make my kids even more miserable from their mom's increasing abuse, under the circumstances, which would likely turn into physical violence, more and more, as the tables started to turn on her.

Besides, I knew that these heinous courts consistently favor the more abusive parent, after all, so fighting that system to get a little justice would be exhausting. I only had just enough energy for survival, when I was that close to X-1. That gal sure knew how to steal my energy! Here, again, I was being a fool. The right course of action would be to accept that man's help and fight her for custody of our children, for their sakes.

That fascinating man had induced the US Navy, (the only lawful executive agency in the county), to threaten to blow the back wall off the jail in the neighboring county, to get someone like me out of custody. The jailers hurriedly let the guy go, at the filthy judge's behest, and all of the charges against him, which were exactly like the charges against me, were dropped. This county didn't have a navy base in it, unfortunately, or I'd have leapt at his offer to help. Quick resolution would mean that I could take care of my own kids again, far enough from their mom.

That ex-cop really knows the law. He still limps from the beating he caught in that jail, from 'His Horror's' assassins. I made sure he got a zapper when I got out, at least.

I just can't eat in captivity—I wasn't actually trying to prove anything. I knew they were watching me closely and would have force-fed me, if I gave them a chance. Those uniformed sociopaths live to exert their will on helpless prisoners. So I sat at the table at every meal with the other prisoners, after accepting a tray of food, then when the jailers weren't watching me any more, I gave away my food. I got pretty popular there, without ever having to bend over. The jailers did manage to poison my water, apparently with a little bit of toxic metal powder fed into the pipe, but the zapper and orgonite took care of that within a few weeks, after I got out.

Those two weeks would have been another kill-shot to my ego, if I hadn't gotten some timely personal guidance from James Hughes in Ashland, Oregon over the previous two years. More about him, in a bit.

My kids' step dad at the time was also a divorcee, and sent all his earned money, along with much of his pension, to his ex-wife and her current husband, so he was feeling the pinch, in the 'musical spouses' game, and apparently wanted to pass the buck along to me, with his wife's active encouragement—nothing personal in it from his end. It's a good thing for my kids that their mom was earning plenty of money, at the time, by doing something she loved, because this john was perpetually on the ropes, financially, and never really adjusted to civilian life.

If you've never been grist for the divorce mill, it's likely that you feel a little condescension or prejudice toward me, by now. I don't wish the experience on you or anyone, for what it's worth, and that was the primer, after all, that set me on my current, more productive life path, so I don't regret it much. There's no getting around the fact that I made a couple of bad judgment calls, by not fighting to get my kids. Doing the right thing is almost always the harder choice.

Note that very few men divorce their wives, and in most cases these men are wealthy jerks, who just want to play house with post-adolescent females, or else they move the mom out of the home, so that they can bugger their own little boys and girls. They can generally afford costly, but supportive lawyers and judges, who are also pedophiles, most likely, and they just want the convenience of not having adult witnesses around.

The courts invariably take their side, of course, and usually tear the kids away from the victimized moms, and bestow them on the abusive or neglectful fathers. If you see kids' pictures on posters these days, it's likely that the Feds are pursuing discarded mothers of conscience, who have rescued their children from abusive fathers, with the help of the new Underground Railroad. Yes, this supportive network really exists, and these courageous women are rarely found, even by predatory CIA psychics.

You won't ever see this reported, on the What To Think Network, because you're supposed to think that their mothers, whose pictures are always shown along with the children's, are criminals.

A couple of years before my divorce, I had the privilege of being able to help put a pedophile in prison. He was a neighbor, whose 'platonic' wife spilled her guts to Melody, my sister-in-law. I confronted him about what his wife had said, and he admitted to the crimes, blaming the little girls and boys in the process—no kidding! The guy was so complacent and arrogant, that he apparently thought I wouldn't do anything with that information.

When a mother in another town accused him of molesting her boy, whom I'd recently seen this guy with in public, (NAMBLA isn't a joke, folks), I provided the testimony that got him sent to prison. The prosecutor and judge were on his side in court, (are you surprised?), and if it weren't for the persistence of the detective, who took my testimony and presented the evidence for the record, this pedophile would have been released by the black-robed pedophile on the other side of the dais.

That was just an average day in a typical county courtroom, though typically pedophiles are set free. The only time they get punished is when the media or an aggressive cop gets involved.

National-Socialism breeds this sort of institutional abuse. The pedophile neighbor had just sold his lucrative business and was on his way to live in Mexico. You might have heard that pedophiles have a much easier life there, entirely free of prosecution.

Until this happened, all the western democratic liberals, whom this guy chummed around with, would tell you that he's the sweetest guy in town. After it happened, many of them visited him in jail to offer him their condolences and told him how sad they were about his 'temporary sickness.' Cultural schizophrenia prevented them from feeling the same sympathy for the large number of little boys and girls, whose lives he had poisoned forever, of course. In spite of what you might have been told, some emotional and psychological wounds never heal. The way courts tend to let pedophiles go free is another one of the calamities of National-Socialism, the relentless destroyer of positive standards of behavior.

I don't keep track of what this fascist bureaucracy claims that I owe them, on behalf of my children, because I just throw the unopened letters away every month. Our zapper distributors occasionally get threatening letters from the State of Washington, but they pay us off the books, so they, too, throw the letters away.

I do know that they're still charging me top dollar for all four kids, even though my older son is 27, my older daughter is 25 and my younger daughter just turned 18. At eighteen, the obligation stops, according to them. I don't cry, 'foul!' to that unlawful agency, because I won't acknowledge their jurisdiction. Also, I'm aiming much higher: the timely destruction of the foreign corporation, which owns and operates all of the courts, bureaucracies and cops in America.

Six years ago, when I committed to destroying the Federal Reserve Corporation, I didn't have an audience. I do now, and I acquired the audience by fighting tyranny effectively, which means I'm on the right track. If you feel that it's time for us to create real government in the world, you can take heart from my continuing success and survival, and maybe that will inspire you to start gifting, if you haven't done so, yet. Genuine empowerment is within easy reach for anyone, now.

I know that many people who gift don't really care about what I'm saying about the family courts, (another oxymoron), and that's good, actually, because it's another bit of evidence that Carol and I didn't start a personality cult or a political movement, both of which are repugnant to us. I have developed a moderate bully pulpit, though, and I wanted to finally get my story on the record, to be read by as many people as possible because,

1) I need to short-circuit a future attempt by my enemies to destroy my reputation.

2) If you, too, have been marginalized by National-Socialism you'll now, perhaps, feel encouraged to finally take some positive action, that will vitiate centralized tyranny.

I know that the sewer rats would like to tell you that there are skeletons in my closet; there's nothing in my closet these days except clothes and a loaded gun. It's better that you hear my personal history in my own words, than their version.

I know something about the magic of the published word, because this network grew out of a few inspiring, substantive reports on the internet. The distribution of orgonite will continue to undermine this tyranny, whether the gifters are aware of the depth of the occult/corporate world order's corruption and tyranny or not. I'm giving plenty of new, usable and empowering information in this book, to people who haven't been tracking our progress on the internet. I hope that everyone who reads this, will feel inspired to make and distribute some orgonite.

This book is my short, concise view of the significance of our discovery, and includes some of our experiences and observations, gleaned during the process of disseminating that information. I assume that anyone who has felt inspired to buy this book can break free of the restrictive protocols of the What To Think Network and overcome the vestigial herd instinct.

Critical, creative thought and open consultation are what will enable us all to find and apply solutions to the outcome of millennia of destruction and exploitation, by the occult/corporate world order. Identifying the problems is a logical first step to finding solutions, of course. Most of the eggs in the What To Think Network's basket ,are designed to prevent us all from thinking critically, and consulting with each other.

They've made distraction into a fine art.

The parasites know, even if most people choose to remain unaware of it, that the victors will be the ones who have generated the most energy. The bad guys' energy is characterized by decay, deception, debt and terror and it's all founded on the 'sucking', fraudulent energy of the mostly-Europoid Pajama People's fears of poverty, reprisal and loss of freedom. Most of this world's prosperity is only experienced by Europoids and a few Asians, after all, which is why genuine teachers and activists have always been the main targets of the What To Think Network. Orgonite's influence is the direct opposite and its creative, life-giving energy is abundant, synergistic and perpetual.

It's not much of a contest really, because our devices absorb and transmute the bad guys' energy, just like turning on a lamp in a dark room stops darkness. The trick to winning is simple distribution.

The scam that keeps the Europoids and prosperous Asian countries in line is the programmed assumption that prosperity is only possible at the behest of the occult/corporate world order. Carol and I are starting to prosper, well outside the boundaries of that paradigm, as are many others, so I think that scam will be put to rest pretty soon, especially if all those free energy device inventors acquire the gonads to share their discoveries. Even the sleepiest PJ person would leap at the chance to stop paying for gasoline, heating oil and electricity.

Positive orgone radiation, which is perpetually produced by orgonite, calms fears, raises hopes, makes the atmosphere and water clear and vital, reverses deserts, generates opportunities for prosperity and creativity, increases harmony, by discouraging predatory and parasitic behavior and not least, helps people have restful sleep, free of nightmares and characterized by lucid dreaming.

Can you see that there's no contest in a struggle between their energy and ours? This is especially true when our technology turns their nasty energy into life force.

Compare your empowering lucid dreams in the presence of orgonite, with the schizophrenic, waking dream-state of the Pajama People that's enabled by their programmed, psychological addiction to the What To Think Network. Your lucid dreaming is just one of the signs of genuine empowerment and progress, and with enough orgonite out there, even the PJ folks will unwittingly acquire some of that!

Some western democratic liberal Europoids, who pretend to be awake, angst over whether the penal system will be privatized. The What To Think Network's easy ability to tie these folks' attention up with non-issues was best demonstrated by their months-long focus on Monica Lewinsky, via droning NPR newsreaders.

The entire government, at all levels, was privatized by FDR's Raw Deal in the 1930s, with the introduction of National-Socialism, under the auspices of a European corporation. All of that inoperable, incurable cancer, that now calls itself the US Government has been a contractual subsidiary of that European-owned company, the Federal Reserve Corporation. Since the mid-30s, nobody in America has owned property, because all of that was handed over by FDR to that corporation, as collateral. I bet you didn't know that. Your property taxes are rent, and if you stop paying those taxes, you'll be evicted from their property, even if you paid off the mortgage.

Americans ARE property now, (probably owned by Beijing, technically), though only on paper, and in the minds of taxpayers and What To Think Network subscribers. Example: you can't get the title to your car or property when it's paid for; you get 'certificates of title.' The corporation retains the title, and if you won't pay unlawful taxes on the property, they're simply taken away from you. Where's the freedom in that? Freedom to 'pay tyrants or be punished'? You and I never had a voice on how revenues are to be collected and spent, unless you have alleged property in a town in New Hampshire. Isn't that taxation without representation? That's what the American Revolution was fought about so, shouldn't that bother you now?

Politically, things are ten times worse, now, than they were then. The main difference between then and now, is that we can get rid of tyranny without a shooting war. We can do it by simply no longer supporting the corporation, which owns the US Government and, allegedly, you, me, our property and our children.

In August 1995, as I was leaving with my dory in tow, after having sold my equipment for pennies on the dollar, I told X-1, in the presence of the cocaine addict who began playing father to my kids at the time of our divorce, the previous October, that she and I were just going to have to take turns supporting the kids. Whenever she wanted me to take a turn, I'd find a way to do it, but that I just wasn't going to send her any money for a while, because I was nearly out of my mind with grief, and couldn't stay near her any more—it drained my energy too much. I told her that I needed to have the energy to earn a livelihood and maybe even re-create my own life, so that I could eventually take my turn and provide a comfortable environment for them. Really, I was in such deep doodoo, because I didn't stand up to her a year before that, and fight for the custody of the kids.

The cokehead, who wasn't even holding up his end in the

'musical spouses' game, eventually sought greener pastures, since it was clear that X-1 would be unsuccessful at extorting me, to be his meal ticket. She had just sold the house and shop and was renting an extravagant home, but right after I left, they moved into a less ostentatious house, because I obviously wasn't going to be paying her rent any more. The home and the big shop behind it, (both were on a landscaped double lot in a nice neighborhood), had developed a lot of equity, so I knew my kids weren't going to miss any meals for a while, and my conscience was clear on that count, at least. Even as an alcoholic, she was mindful of her purse strings, which is probably why my replacement had been stealing my kids toys, videogames and recreation equipment and selling it all to get money for his own drugs. I found out about that much later. I'm glad I didn't know that at the time, because I don't think I could have been as detached.

I was finally economically stable enough to resume my responsibilities in the fall of 2001, seven years after the divorce, but I was still unable to bear the thought of sending X-1 money, (her business was doing very well, thankfully), because it still felt like extortion to me, and I was forbidden to visit them by the corporation.

Based on her past performance, and the fact that my kids weren't lacking anything, I felt pretty sure that none of that money would go to the kids. It burned to think that I would essentially be rewarding her, for having trashed my life and abducting our kids, with the corporation's blessing.

So, I rather sent my older daughter a thousand dollars a month during that time, so that she wouldn't have to work when she was in college. She was at Brandeis University on a scholarship. It's a tough school, so having a job on the side would have meant less time for studying. When Bevin graduated from that Boston school in May 2003, I bought her a used car and gave her as much startup money as I could, and she's done very well since then.

I gifted X-1's house and neighborhood in Mt Vernon, Washington, in May 2004. They had moved back to that town from nearby Anacortes, (where I was arrested in January 2001), after she discarded her second husband. The cops had been tipped off by someone and were trying to find me that day, so I played cat and mouse with them, during that neighborhood and high-school gifting sortie, which was kind of fun. My second daughter, Nora, was going to school there, at the time and it was only a couple of blocks from the house.

I think the Feds tipped the cops off, because I'm pretty sure X-1 didn't know I was around that day, because she was far off, with her new, wealthy boyfriend. By then, I no longer had a driver license, so I was easy meat for cops. I'd thoroughly gifted the fortified court/cop-shop/jail complex in town two years before that. That corporate fortress was my temporary home, in January 2001.

X-1 moved out of that house a couple of weeks later, onto the summer estate of the guy who owns the nuke plants. They lived in Seattle together, until he married her, then he built her a mansion on a mountain he owns, in a tourist area.

I sure didn't expect such immediate, dramatic results, from that gifting sortie, but it's good to never underestimate orgonite's potential, of course. I was especially glad, because after that I could visit my kids anytime, after they let me know whether she'd be visiting them or not.

When the mom moved out, our very responsible 24-year old son, Arian, moved into the house to take custody of the two teens. I've been sending them twelve hundred a month since then, and until Carol and I moved to Florida last fall, I visited them as often as possible—a pleasant, scenic, four hundred mile drive from our home, in Idaho. I've always loved the Puget Sound area, and it's too bad that I can't go there, without risking corporate imprisonment.

Every time I went to see them, I risked going to jail for a month. That judge really didn't like me, especially since one of his would-be victims offered to help me beat that rat in his own nest. It's curious how sanctimonious a black-robed pedophile like that judge can be, in his own courtroom, backed up by a parasitic corporation that only owns debt, (represented by that gold-fringed admiralty flag), and rules America by extortion rather than by law.

My two teens chose to come live with Carol and I for a few months, in the summer and fall of 2003, when their mom was still at home, but they didn't care for Idaho much, so they moved back to Mt Vernon.

Right now, I'm spending $1400 per month on the teens, one of whom, Nora, just turned eighteen and is enrolled in a prestigious haircutting school in Seattle. She loves the experience and hard work, and that school guarantees a high paying job, after a year of training.

The mom is generously paying her tuition and the rent for her little apartment. She spent a thousand bucks on nice furnishings for her, too, but is leaving it up to Nora, (me), to take care of her other expenses for now. Nora got a part time job, so we'll play it by ear. I'm sending her a hundred per week until I'm sure she's comfortable and has everything she needs.

Nora's the one who was my psychic lab assistant, when she was nine and twelve years old, and I was making and experimenting with orgone devices, but still living in my 1980

Honda Civic, (the seats on those cars recline enough so that a tall guy like me can stretch out properly). I was using a friend's cabinet shop, early on, and had a tiny apartment in Ashland, Oregon when she chose to come live with me for a few months, at age 11.

That move was against her mom's wishes, but someone had kindly let Nora know that she was then old enough to choose which parent to live with. It was when she was in Ashland with me, that I learned about orgonite and started experimenting with it, with Nora's timely help.

Nora had gone through death at age five. She had ingested a visiting friend's stash of Nutrasweet, which put her into a long-lasting, mild epileptic seizure. I was at work at the time, so her mom frantically took her to the hospital, where an incompetent doctor injected so much poison into her, that it stopped her heart. She was airlifted to an even worse hospital in Seattle, (I think 'Children's Hospital' is called that because of the median age of the doctors there), and revived, though she stayed in a coma for a couple of days. We were sitting with her, when she came out of the coma—it was cute; she opened her eyes and leaped into my arms—and we quickly abducted her from that Gulag, immediately followed down the hall by a little female doctor's noisy and threatening protests.

X-1 became lucid for a few days after Nora's recovery, thank God. If I'd been home at the time, I'm sure I could have convinced X-1 that our nearby chiropractor friend's competent help would have been a wiser option, and that he would have quickly stopped the seizure. We later learned that an icepack on the head stops those aspartame-induced seizures.

The little serial killers, (MDs), while Nora was still out cold, had tried to coerce us to put her on dopamine. Yikes. She had fewer and fewer seizures, until they stopped about a year later. The ice packs always pulled her right out of them. We didn't pay the $17,000 that the hospital charged us for that fiasco, of course. That would be like paying for the bullets that the National-Socialists used to shoot your grandfather.

Going through death made my daughter intensely psychic, just like Danion Brinkley describes, but she never spoke to me about it until four years later. Nora had been relentlessly hammered by her mom, every time she mentioned a psychic observation, which is why she almost immediately stopped mentioning it to everyone else. When I returned to the area, when Nora was nine, she described the blue energy that was coming out of one of my new orgone accumulators, and I immediately recognized and was deeply grateful for her gift. That was the first time anyone honored her gift, so she spilled her guts after that, and we were close partners in research. Her classmates and teachers at school had hammered her as relentlessly for being psychic, as her mom did. Mt Vernon isn't a fun town to live in. When she was in Ashland, Oregon, with me later on, the school kids and even her teacher were supportive, as were James Hughes and Dorothy West. That's a pretty cool town.

I'm continuing to pay my older son, so that his sixteen-year-old brother, Cameron, can have a nice home. Their mom sends them money every month, too, and it's kind of nice to have a quasi-cooperative arrangement with her by now, though if she ever saw me she'd want to get me incarcerated. Number One Son works full time. I give my younger son, Cameron, two hundred a month, which he saves, because he's also earning some money creating commercial web pages with Nora's nice former boyfriend, Decker, who stayed behind in Bellingham, when she enrolled in haircutting school in Seattle. I feel very proud of Nora and Bevin for having stable relationships with their guys. What are the odds?

I send them postal money orders and sometimes deposit money in their bank accounts, but won't keep records because it's nobody's damn business, but ours. My kids would all back me up, if it came to that, which probably won't happen. They've all got consciences.

I was their dad and also their default mom in their formative years, because their biological mother went into daylong, often-violent psychotic rampages upon waking, most days. Imagine living twenty-two years of that, with very few lucid breaks. Looking back, I can see that it was a lot like taking care of someone suffering from Alzheimer's, though now I realize that she could have behaved differently, if she'd chosen to. She threatened suicide once a week or so, usually in front of the kids, each of whom she stridently blamed for her unhappiness, when I wasn't around to blame. When I was still in the home, they didn't take that to heart much, because I reassured them that they each rather made me happy by just being alive. Fortunately, most of the younger kids' formative years passed while I worked out of the home, and they spent more time in the shop with me, than in the house with their mom. I took a lot of breaks to play with them and to take them on outings.

It was X-2's brief periods of relative calm and lucidity that kept me hoping that she was going to mellow out someday, but by now I understand that people always choose whether to behave like predators or not, in a given moment. We're all thoroughly programmed by the What To Think Network and psychology professors, to assume that predators are somehow just victims of circumstance. Strange culture... by extension we're also deeply programmed to ignore, wink at or even enable the dynamics of the predatory behavior of politicians and corporate executives.

I hope to be making enough money, by the time my youngest

finishes high school, to pay for art school for him. I also want to help my older daughter go to grad school at Harvard, when she's ready, though her current employer will likely foot the bill. She lives and works in Cambridge, Massachusetts. Like any parent might, I'd love to buy them all houses someday. Haven't you noticed how National-Socialism rather made the Europoid Depression Babies, my parents' generation, into niggardly wealth addicts? Not many of them help their kids much, because they were spiritually poisoned by the National-Socialist dole in childhood. Their own parents did everything they could to make the Depression Babies' lives better.

In America you commonly see Depression Babies with million dollar homes, $200,000 RVs, a new Cadillac every year, and bloated stock portfolios, who won't help their children and grandchildren with groceries, when they're in poverty and working their butts off.

I once met an Alaskan Indian woman, who had never been around whites until she went to Bellingham, Washington, to go to high school. In her culture, the old folks were treated with respect, because they were generally wise, kind and generous. She experienced severe culture shock, when she encountered National-Socialist Europoid Depression Babies' obnoxious, noisy, schizophrenic and self-gratifying behavior.

I think the peak of National-Socialist surreality, for me, happened when Carol and I had rented a campsite at the RV Park, in Ft Pierce, Florida, in November 2000. We were coming back from a restaurant at around 10PM and the road to our campsite was blocked by hundreds of Europoid Depression Babies doing the hokey-pokey in a conga line. It looked and felt, to Carol and I, like a staged, seething National-Socialist rally in 1935 Berlin, and we were incredulous. Yet another initiation for me.

We enjoyed hanging out with some of the Quebecois old timers at that campground, at least, who were openly friendly and curious, instead of paranoid and prejudiced, as was the case with the hokey-pokey aficionados.

A nice development of being cast out of my home, years ago, is that I'm making more money by now, than I'd probably ever have made as a sign painter, and it feels good to finally be able to help my kids. I wish I'd gotten this kind of help, when I was starting my own life. Most Depression Babies, my parents' generation, are curiously unable to give anything away, and they reflexively invoke the [Really!] Great Depression, whenever their children or grandchildren express a material need. It's a good sign that my generation and the following ones are less niggardly.

When I was in Uganda, I met and talked a lot of people, across several generations on our gifting sorties and I got a chance to closely observe how my Ugandan hosts related to their own families and cohorts. I noticed that everyone in extended families helped each other materially, as much as possible; there are no beggars on the street, even in the capital, and nobody is starving, though the country is quite poor. I noticed that the devastation, bloodshed and starvation that followed 'independence' for a generation or so there, didn't cause anyone to become stingy or self-seeking, the way my parents' generation are, on account of National-Socialism's birth, using the Great Depression as an excuse for their infantile behavior.

Ugandans also don't have any social programs, other than the token but mostly sincere efforts by a few foreign governments and charity organizations.

If I get rich before the old world order collapses, I'll pay whatever amount of blood money is necessary to get the

Federal Reserve Corporation off my back, but only if it's paid to X-1 directly and she then calls off that black-robed pedophile and his leg-breakers. She can do, anytime, with a pen-stroke, of course. If the Federal Reserve Corporation is destroyed before that comes to pass, I'll just use that money to help my kids, instead.

I'm only willing to pay blood money, because she'd then be less able to psychologically torment our kids. Whenever they see me, if she finds out about it, they pay a price in the form of their mother's enraged torment and abuse, if she's off any of her meds.

I can go on living and prospering outside the system, otherwise, and I can go anywhere I want with my World Citizen passport. Of course, all bets are off, if the bad guys just decide to shoot me in the head, or toss me in a prison or mental hospital, (the 'thorazine shuffle' is a new phenomenon among political prisoners on America, which will also puzzle future historians), but where's the fun in life, if there's no risk?

I wonder why so many people just cave in to the National-Socialist corporate parasites. The PJ folks' bending over and grabbing their ankles, en masse, for the Federal Reserve Corporation—the ultimate Bubba—is the only thing that keeps those parasites fat and contented, and the masses enslaved; that and the Red Cross blood-banks.

You might still assume that I'm just a malcontent, if you've read this far and I've offended your sensibilities, but if that's how I've impressed you, how do you explain away your own deep-seated fear of running afoul of this admittedly parasitic system, and your feeling of powerlessness in the presence of this corporation's clean-head, black-uniformed leg-breakers? Do you really think that's the proper way to live in this world? Do you honestly believe that you're free?

I remember when people, at least Europoids, in America, felt comforted by the sight of police rather than intimidated. That was before Depression-Baby-enabled, wholesale slaughter got started by the corporation in Vietnam.

Are you as genuinely happy as I am? I don't think real happiness is possible, when one is attached to an illusion of security. My false sense of security vanished in a moment, when I stupidly lost my children to the lawless American court system on their mother's whim, but the experience forced me to either recreate my life as a sovereign person, or go the way of the new hobo class and endure a meaningless existence as one of the faceless, disenfranchised, playing musical spouses and living at subsistence level, at the whim of yet another rudderless female.

I know that most white people in America call themselves free, while they're really just voluntary corporate slaves; voluntary taxpayers. They're afraid of the new, omnipresent, black-suited, jackbooted cops in body armor but they can't put a finger on why that is. This fear is schizoid, frankly, so it usually translates as racial prejudice, even in suburban towns like mine where there's no violent crime, but the cops are omnipresent, aggressive and decidedly 'apart' from everyone, but donut vendors.

The automatic result of destroying this poisonous corporation—thus returning political and economic power to the states and communities, as nature intended—isn't anarchy; it's freedom and prosperity. Reflect on that, then observe the PJ folks as they numbly genuflect to this treasonous, parasitic National-Socialist regime and their horrid, jackbooted, KGB/STASI-led Homeland Security Abomination.

The trick is to get the deed done, without violence, before this Nazi agency is instructed to start openly enforcing the treasonous Patriot Acts. In that case there really would be a brief shooting war, throughout the land and I sure don't want to see that happen, do you?

Then, while you're still rightly angry about this terminal political cancer, make some orgonite and deftly distribute it around the perimeter of the fountainhead of corporate rule in your community, the fortified courthouse/cop-shop/jailhouse compound. If you're particularly resourceful, maybe you can hide an orgonite muffin under a black-robed pedophile's own bench.

Orgonite gets through metal detectors, undetected, every time.

It would be a hoot to be in that predator's courtroom/rat-nest after that. The dirtier the predator, the more unhinged he gets, when orgonite is close to him, as a rule. Later, I'll tell you how Carol took down a notoriously predatory and felonious county prosecutor, in his own courtroom, barehanded and from twenty feet away. She made history that day, a year or so ago, and many folks have been replicating those happy results since then.

Have you noticed that there have been no 'terrorist attacks', since the Feds blew up the WTC, almost five years ago? I agree that the demolition of the levees in New Orleans initiated a blatant, federal terrorist attack on the city's black population, but in that context they blamed 'nature,' not swarthy men from the Middle East. Even so, the increasing talk about 'terrorism's threat', on the What To Think Network, has become more and more strident, since the feds blew up the World Trade Center. In fact, by now all of that is just like the 'doublespeak' you read about in the book, 1984, when you were a kid.

What's not so funny is that Congress and the fake President have been generating legislation that has exceeded even Hitler's and Stalin's bloodstained fiats. Thank God they haven't dared try to enforce most of that! If American's weren't armed, all of it would be in force, and those waiting concentration camps and guillotine chambers would be engorged by now.

Hitler had to get the Germans to voluntarily disarm, before he could institute his draconian measures successfully. He did it by burning the parliament building and blaming terrorists, (commies in that case), for it. Notice that people, except the expiring Depression Babies, are no longer as stupid and gullible as that generation was. Fortunately for the rest of us, the Depression Babies' numbers are so decimated by cancer, Alzheimer's, strokes, heart failure, arthritis, diabetes and alcoholism, that they're no longer a potent political force, as they had been in their prime—during the Cold War and Vietnam Conflict. Now, they're mostly so firmly attached to Medicare's poisonous teat, that I doubt they'll be heard from, politically, again.

That's poetic justice, as far as I'm concerned, and it's recompense for their blind, schizophrenic devotion to serial killers, (MDs), within the parasitic National-Socialist paradigm.

I assume you realize that you're not 'safer' because the Feds increased the power of the federal government, after they blew up the WTC, and openly instituted even more intrusive surveillance and monitoring of all Americans. The new airport Gestapo make everyone feel uneasy, even the Europoid Pajama People and Depression Babies. Have you noticed? People are avoiding air travel in droves, because they don't want to be publicly humiliated at airports. The airlines can't afford to feed passengers on cross-country flights any more, due to critical loss of revenue; so many passengers are bringing coolers as carry-on luggage.

The 'deadbeat dad' issue generates almost as much raw emotion and summary judgment, among PJ folks, as 'abortion,' 'skinhead' and 'pedophile' do.

You could be a saint, a hero, the composer of a symphony, or the savior of your people, but if slanderers' cacophony leaves you wearing an unflattering label, that will be all that most people will remember about you. If you try to defend yourself under those circumstances, it would be just like answering loaded questions, such as, 'Why do you beat your wife?' or, 'Did your mother ever catch you masturbating in the closet?" The What To Think Network will do their best to guarantee that you'll be seen as a pariah, if you're a genuine pioneer, who won't be owned and controlled by the occult/corporate order. This is how the What To Think Network quickly gets rid of honest politicians. There aren't any more honest, national politicians. See how thorough they were?

In a worldly sense I'm not remarkable or particularly gifted, but I've earned a good reputation, through honest means, and I wish to keep it and hopefully improve it, so maybe a stitch in time saves nine, in terms of preserving my reputation from media slander, in coming days.

I told X-1, nine years ago, that I would make her famous, but that I'd never take her fishing. That was after she produced a bowlful of freshly-killed, intestinal worms. They came out of her during a colonic irrigation, while she was zapping for the first time.

I keep a photo of some of those worms on the homepage of World Without Parasites. I got the photo from her, after her first zapping session, during a self-administered colonic irrigation. She had been ingesting anti-parasite herbs—the finest on the market—for several months, without producing any visible results. Her business was giving colonic irrigations to people, and she was happy in her work and thriving, as I mentioned.

When I asked X-1 to do that zapping experiment she said, 'Nothing you make could be any good!' but she zapped, anyway, and I have a fine public record of the results, that's been helping me sell a whole lot of zappers, over the years. Dr Hulda Clark has claimed that zapping won't kill intestinal worms, but even before that experiment I had become convinced, by consistent evidence and testimonials from many customers, that just using a zapper around the clock for a few days kills them all.

X-1 is still trying to give me the business in a similar way, but it's not working out for her, because I won't play by the corporation's rules. Carol and I were married during our travels, in a county courthouse, where social security numbers still weren't required for the record, so the Federal Reserve Corporation, therefore this National-Socialist regime, considers us unmarried. They hate to even hear the word, 'common law,' even in the context of marriage, so I'm not worried about them invoking that. It's just like the way MDs in the West mostly hate to hear the word, 'parasite.' In that case, they are parasites, of course, and you apparently hurt their feelings when you say the word.

It's all about the slave number with the corporate/government parasites. They can't escape that narrow path, except for when they get frustrated and have one or another of their federal police agencies just blow someone away or spontaneously throw them into prison or a nuthouse—as has been the case with many freedom activists lately.

Scholarly, old Irwin Schiff and his wheelchair were recently tossed into prison, just because he proved, beyond question, that the federal income tax is entirely voluntary. Elderly Roger Elvick was thrown in a nuthouse and forced to do the thorazine shuffle and, until recently, (after Todd Platzer gifted the nuthouse in Cleveland, where Roger is being warehoused), was given daily injections of especially debilitating psychotropics, to prevent him from even speaking to people. He had helped a lot of people like me in the courtroom, over the years, and the sewer rats, unable to defeat him in court, simply gave up trying, arrested him for nothing an locked him away. Did you believe that this horrific treatment of innocents stopped, when Communism fell in Russia?

Tormenting political prisoners in America isn't very patriotic of the courts, Gulag Archipelago and the Homeland Security Abomination, don't you agree?

Most of the high profile personal sovereignty advocates are incarcerated, or have been murdered by now, so if you meet one who says he's 'beating the system,' run the other way, because he's likely there to get you profiled by one or another of the alphabet soup agencies as an activist or potential activist. One of them almost got me disappeared by cops, in Las Vegas a couple of summers ago.

As I said, this parasitic system is irretrievable and has to be replaced with a workable government. Nobody beats this system, otherwise. If you want to be an activist, do something for which the sewer rat agencies won't dare come after you, at least for now—make some orgonite and start taking territory away from them!

If the Fed considered Carol and I married, X-1 would automatically own everything that Carol pays for, and the Gestapo would take it all away and give it to her. All of my purchases take place without involving the slave number, of course, but Carol still uses a credit card and gets bank loans, which means her major purchases involve the slave number. I know that most PJ folks wouldn't balk even if their slave numbers were painfully branded on their wrists and foreheads. You can bet the Homeland Security Abomination cowboys would enjoy that nationwide rodeo roundup a lot, if 'the mark of the Beast' brand, (implant), on the body became literally mandatory.

Enough orgonite in the PJ folks' neighborhoods will probably make them stop fearing this alleged government's displeasure so much, and that means they will stop supporting it, too. They're not stupid, after all; they're just asleep.

I really miss being in Africa. Over there, most of the people you meet want orgonite and they start using it immediately and intelligently, to improve their lives and their environment. Over here, the only way you can get a foot in the mentally-conditioned Europoid PJ folks' doors is if you're scamming them like the fake government and the What To Think Network does. They're programmed to numbly 'go along to get along', and even to voluntarily pay for their own enslavement.

Pajama People, until now, have been a masterpiece of external mind and behavior control, and self-policing. My only fear is that if they wake up too quickly, they're going to start shooting, and there really would be chaos then. They're too unaware to know who the logical targets are, if shooting broke out. In the best case scenario, we few genuine activists will have undermined and weakened the parasitic order so much, with gifting campaigns and etheric predator safaris, that the corporation will have simply lost their base of support. Their sole support is the programmed terror of the PJ people of losing their 'security,' (read: comfort).

After we do enough gifting, the PJ folks might just continue to be relatively comfortable, sleepy and complacent, as the corporate order dissolves and is replaced by the democratically-derived rule of law. The rest of us can create a fabulous world, without risking murder and imprisonment, at the hands of the parasites' wise guys. I don't call PJ folks 'sheeple', because that implies judgment of our fellows, which we're not capable of. I direct attention to their programmed behavior, instead, because they can wake up and change that, in a heartbeat if they choose.

The break with X-2 in April 1997 was a lot cleaner—she simply wanted nothing more to do with me at one point, so I moved back across the US in May, to be a little closer to my kids. That was two and a half years before X-1 sent the corporate hyenas after me.

Though I was still not able to do much more than feed myself, X-1 let the kids spend time with me, during the following summer, which is when Nora first started helping me with my research. I think X-1 was feeling pretty good about my failed attempt to remarry, so she eased up; also, her business was doing well and she probably enjoyed seeing me live in penury, with no love life, while I slept in my friend's cabinet shop, outside of town. That was when she magnanimously agreed to use a zapper once.

My energy steadily dropped, from being that close to X-1, even though she wasn't being particularly malicious at the time, and nothing I did seemed to raise my vitality, so I moved to Oregon, one state away, to try, again, to rebuild my life, rather than just survive. That was a good move. I certainly wasn't in a condition to start a romance, because I didn't want to be financially beholden to a female, nor did I want to burden anyone with the emotional X-1 baggage, quite yet.

I think that X-2 was a typical Monarch-programmed honey pot, thrown at me to put the brakes on my development, on account of the rising profile I was already starting to develop with the zappers. Nearly everything I was passionate about terrified that poor gal, so of course I stopped developing myself 'for her sake.' According to D Bradley, anyone who is destined to do something remarkable, in the way of liberating humanity from tyranny or healing the world, is pegged at birth or shortly after that, by the dark masters, personally, and plans are made, then, to either kill the child or ruin his or her life through Monarch programming. I've had enough run-ins with old, dirty magic that it seems entirely feasible to me and, in fact, explains some phenomena that would otherwise make no sense. I'm not a flake or ideologue, so I have no vested interest in whether this is true or not. I live my life as though it were true, and I get some interesting confirmations along the way, most of which I share publicly.

The difficulty in relating to Pajama People, if one has had preternatural experiences, is that they have no personal frame of reference for anything that's not sanctioned by the What To Think Network. For a year before the felonious feds got in our way, Carol and I had contended with parasitic and predatory aliens more than a few times, on account of our research, so when I later read Dr Reich's account of the spaceships that were targeting his lab, I had no problem accepting his account at face value. If I'd read that before I had experience with unpleasant off-worlders I'd have had to put all that in my 'wait and see' file, which is what I hope you're doing right now, with my supernormal accounts, if you haven't had direct experience with things like this.

We met Al Bielek at the tail end of our months-long travail, at the hands (sic) of those aliens, so I was able to absorb most of his accounts without discomfort, especially since what he was describing about his experiences with them was pretty much like what we, ourselves, had been experiencing. I'm afraid that until these things become 'common sense,' they'll probably remain outside the considered realm of possibility for The What To Think Network's subscribers.

I personally believe that X-1 was tossed in my path in 1971, when I was 22 and had started to develop myself again, (I did it, briefly, when I was 19 but was funneled into the US Army then), because I'd been cast out of MK Ultra at age fifteen, on account of 'character issues,' which apparently means that I had a character, therefore was unusable to my handlers and programmers. Most people choose to have no character, because personal integrity induces us to make the hard, (right), choices more often than not, (like fighting for custody of my kids, for instance, instead of rolling over), and that's often uncomfortable and risky. A lack of character is what makes dissociative programming 'take' so well, in countless millions of Europoid and Asian Monarch and Tavistock assets.

If your red flag of incredulity has popped up just now, please bear with me because I'm only giving you a subjective account. For what it's worth, I've seen so many parallels in the personal lives of people in our network, that I no longer question whether this stuff is valid, even disregarding consistent, firsthand accounts from people like Al Bielek and D Bradley. Enough reading of documented accounts by several reputable authors of the Monarch Program and Tavistock Institute, enabled me to accept the possibility of this massive scale horror in the first place, years before I knew about orgone.

Rather than simply continue to liquidate the countless thousands of boys and girls who were unusable to the CIA, British MI5, KGB, Mossad and other secret police agencies on account of character issues, I think it eventually became prudent for them to sideline us all into dead-end lives with overwhelming, actively reinforced, self-limiting or suicidal, post-hypnotic suggestion, along with some timely 'guidance' into parasitic personal relationships, with their own active, but schizophrenic personnel assets. I think we can get a picture of which nations are most heavily populated by Monarch assets, by looking at suicide statistics, in fact. Suicide is almost unheard of in Africa, where there's no mind control, and before World War II it was also mostly unheard of in the countries where suicides reached epidemic proportions, in recent decades.

I see the footprint of mind control history over and over again, in correspondence with people who come into the gifting network, and the psychics assure me that I had been buggered for years, in the CIA's MK Ultra from the time I was four, along with millions of other boys in that particular Monarch program. I'm not prepared to claim that my history with them is factual, as I said, but I'm convinced it's true, at least. X-1's inexplicable, daily schizophrenic, predatory rages and X-2's paranoia and her deep-seated desire for corporate security, rather than economic freedom, certainly also fit into the category of dissociative programming, I think. These things go a lot deeper than the programming PJ folks get every day, from the What To Think Network.

As the internet history of the gifting network shows, an awful lot of 'guided missiles' were aimed at the discussion boards in the earlier years; people who were obviously not completely aware of their actions, but who did everything possible to disrupt the postings and generate chaos. The substantive content of our first three internet boards, most of which was deleted by the webmasters, as soon as I left those boards, would need to be retrieved in text-only form from internet archives. All that remains of the more easily available records, are the postings of crackpots and disinformants, unfortunately—see the game?

My own board, Etheric Warriors, was destroyed twice, last fall, by NSA hackers and thousands of substantive reports and posted photos were essentially lost. I say, 'essentially', because in practical terms almost nobody will make the effort to drag text from archived HTML files, no matter how informative it is. Jacques Lasselle, our very capable webmaster, has been making backups every day so, that the NSA will be less inclined to make that obvious effort again.

I fell for X-2 when she crossed my path in June 1996, and expressed interest in me because she was gorgeous, witty and available, and because I was a lonely bonehead. I did get a good introduction to subtle energy dynamics, during my year with her, though, because she was terrific at making and using flower essences, and was keenly energy-sensitive. I helped her develop and market her peerless products and they were selling like crazy, in several reputable stores, mostly by word of mouth, when she suddenly panicked and pulled them all off the shelves, in the three states where we were selling them.

Flower essences are as harmless as zappers are, so there was no real risk involved in marketing them, of course. She'd had a few years experience producing her flower essences for an established company, before I met her, so I know there was no basis for her sudden terror. She was contemptuously exploited by the parasitic, self-seeking and duplicitous new-ager who owned the Flower Essence company, but she saw herself as a slave and victim, so she just accepted that treatment. The grinding momentum of corporate slavery's self-policing programming is what keeps PJ folks—even the ones like her who believe they're awake—in deep slumber, after all. There's an innate duplicity in slave consciousness, though—a love/hate attitude toward their masters that's as inherently schizoid, as many other aspects of Western culture are.

I had convinced X-2 to quit her job to develop the line of remedies, because I was making enough money with sign painting and selling zappers to modestly support us both, after a few months. I was in one of the 'feast' periods of the feast/famine sign trade at the time. I knew from experience, that developing a business in a timely way requires that one's first efforts will be applied to the venture. Working all day for someone else doesn't leave you much energy for yourself. She thrived in the corporate sector at the nuke plant, aside from regular, paid 'timeouts' to recover from the serious illnesses and occasional major surgeries that the DOR in the nuke plant caused her, but working for private bosses for less pay was grinding her down.

When she became too terrified to participate in the open market any more, she went back to being an exploited wage slave in the new-age/herb shop where I met her. She also started 'shopping' for a more lucrative, but less challenging mate then, during weekend-long absences.

One of her favorite jokes, (I should have taken the hint) was this one:

A wealthy, middle-aged businessman came home to his pretty, young trophy wife one day and said, 'I'm ruined—I lost everything! You still love me, don't you?' She said, 'No, but I'll never forget you!'

The end of her business, was the beginning of the quick end of our relationship, cemented by me being openly cuckolded.... again. Bonehead.

After two and a half intensely lonely, but very productive years in Ashland, Oregon, with a hiatus in Anacortes, WA, and the pokey, Carol and I got together in Idaho in June 2000. I was finally able, at age 51, to be in a loving relationship with an attractive woman, who appreciates the value of commitment, as I always have. If you're finally starting a relationship with someone who's genuinely committed to you, consider that you'll probably need to adjust your understanding of the dynamics of mutual love and commitment, because you're no longer a potential victim. It's not as easy as it seems, to make that adjustment, but it's well worth the effort. Personal commitment to a marriage is a rare commodity right now in the West, due to the spiritual drunkenness that resulted from National-Socialism's intrusion into everyone's bedrooms, a few generations ago.

Female Europoids are programmed to be self-seeking princesses and also, (often dissociatively), programmed to see themselves as chattel. Europoid males are programmed to assume that they need to revere females, and also that it's okay to cheat on them. Anywhere we look in Western culture, which is erroneously called, 'civilization,' we find deeply programmed schizophrenia on a par with the Santa Claus scam and Musical Spouses.

I've come to believe that the root of the problem is patriarchy, which established itself in human society in Babylon, if I'm not mistaken, and perhaps also China at the same time. It was in that the land of Babel, during the time of that culture, that the traditional, neo-matriarchal values of Eurasia were suddenly reversed. It happened much later in Northern and Celtic Europe, when Rome eventually subverted all of those cultures. It happened much, much later in the Americans and later than that in Africa, where the shift is still not a done deal. Most of Asia turned to patriarchy earlier than Europe did.

I just got an interesting confirmation from reading French historian, Jean Markale's, scholarly and thorough, "MONTSEGUR AND THE MYSTERY OF THE CATHARS". Markale has specialized in the study of pre-Roman Celtic and other European cultures and the universal theme in Europe and South Asia, before Rome's conquest, was that the sun was represented by a feminine deity, and the moon was represented by a male deity. Patriarchy is based on the opposite, and the switch was apparently made from Babylon, outward to the Indian and European cultures.

Piggybacking that switch to patriarchy came the concept of dualism, which programs people to believe that physical creation is intrinsically evil and even that women are corrupters of men. These ideological prejudices were absent from pre-Roman European and pre-Babylonian Near Eastern and Indian cultures, and dualism spawned devil worship and the general belief in evil, as a distinct force or entity. The schizophrenia inherent in churchianity, is an outgrowth of the belief in dualism and devils, I believe, which came into Christianity through manipulative Roman/Persian belief paradigms and clergy, not through Judaism, (Jesus' own religion), which has no significant devil doctrines or dualism precepts.

Another foundational theme of patriarchy is the assumption that men are primarily mental and women are primarily emotional. Dorothy West, the old druid witch, who took me under her wing in Ashland, Oregon, told me that according to the Doran Order, which she and her forebears had been members of, for over three thousand years, (therefore is pre-Roman), men are primarily emotional and women are primarily mental.

I feel that I've come upon the root—patriarchy—from which culturally programmed schizophrenia and slavery grew, in Babylonian times, and eventually spread through most of the world. I wonder what I'll be able to accomplish with that little bit of information. It certainly holds up under scrutiny, and so much of history that made no sense before now, makes sense to me, in light of this.

It's a genuine miracle, when a Europoid man and a Europoid woman develop a healthy marriage, and when you toss in the fact that National-Socialism has insinuated itself into all marriages, through the parasitic and intrusive rule of corporations, the odds are even greater that their marriage will fail. Pile onto that burden the incessant What To Think Network pressure to turn romance into a religion.

I know that 'returning to the way things were' is no answer, because the past is the foundation of mass schizophrenia, after all, and in the days before National-Socialism, churches, rather than Big Brother, insinuated themselves into private marriages, forbidding divorce and enabling incest and other social plagues among the schizoid PJ folks in the pews. Racial and gender prejudice were considered healthy in those days, too. The benighted Europoid generation in the 1930s welcomed National-Socialism into their workplaces, parlors and bedrooms, after all, under the auspices of the Federal Reserve Corporation and its European equivalents: Hitler, The London School of Economics, Franco and Mussolini. The French people got sucker punched, yet again, in the process, too.

I have a lot of faith in humanity, with the caveat that only a small, proactive percentage of people will ever take the initiative to improve our world. The vast majority, who wish to remain asleep and reactive, would cut off an arm, before they'd consider rocking the boat, so they're not really a threat or a benefit to the world—just voluntary cogs in any prevailing machine.

I think it's prudent to be courteous to the Pajama People, and to genuinely love them, but also to hold them, at arm's length, in order to maintain a sanguine disposition. If you seek validation from 'the masses', you're going to make yourself crazy, because instead of getting the PJ folks' gratitude and understanding you'll set yourself up for savage ridicule and ostracism. They get that way, whenever their dreamy, What To Think Network opiate paradigm is threatened—haven't you that noticed by now? So, be decisive! Stop playing those stupid 'devil's advocate' self-policing games and just get busy. The answers to the Big Questions will sort themselves out along the way, but you need to develop momentum now.

Get personally involved with the rest of the proactive few, around our lovely planet, as Carol and I are doing. You'll develop the same faith in humanity that I'm enjoying, and you'll make some genuine friends in the process, who will stand by you, rather than stab you in the back or drain your energy.

In the best case, the Federal Reserve Corporation and it's interlocking European and Asian National-Socialist affiliates will be undermined and eventually destroyed, of course, and everyone will suddenly have a better life, especially those millions of unjustly imprisoned Americans who will gain their physical freedom. Since America was to have been the keystone of these declining globalist corporations' tired and discredited 'Novus Ordo Seclorum,' our victory in America will probably be a major victory for the entire human race.

Orgonite tends to separate the sheep from the goats, though a few goats make orgonite these days. Many PJ folks call themselves activists, as you can see from the weird phenomenon of NPR, which is the most depressing, oppressive organ of the What To Think Network, but to which an entire generation, (mine), genuflects almost every morning and evening. This generation are the bitter, cynical environmentalists, feminists, frayed Marxists and PC mavens and others, who feel that the only way to improve the world is for National-Socialist governments to create and enforce draconian measures to ensure that everyone else will think, talk and behave as these western democratic liberals see fit.

As a rule, (and every rule has exceptions), when I meet someone who calls him/herself an activist, but dismisses orgonite out of hand, I simply leave them to themselves. I put my energy into helping the few real activists who will make orgonite, and use it productively. My attitude may seem simplistic to you, but if you look closely, you'll find that the gifting network is incredibly diverse and mostly made up of sober, committed people, who generally get along very well with each other, regardless of personal, cultural, religious and racial differences. Their common purpose is healing the world and they get confirmations for their efforts, so they naturally want to share their stories. The operatives of the occult/ corporate world order stand squarely in the path of anyone who commits to healing, so dealing with them appropriately is a necessary inconvenience for now; ignoring them guarantees personal failure, if you're a genuine activist, because they're guaranteed to find a way to exploit your weaknesses or those of the people closest to you.

'Activists' of the receding paradigm, on the other hand, seek advantages over each other, and constantly wrangle over organization, funding and even ideology. They put their hopes into influencing corrupt government officials and inciting ersatz-noble confrontations between 'the people' and corporations, as an excuse to get even more draconian National-Socialist legislation passed. The corporations pretend to be patsies, but they're the ones who are controlling the very politicians, whom these benighted fanatics assume will put the corporations in their places with more and more legislation, which simply makes federal bureaucracies increasingly powerful and intrusive.

Most of the top people in the hierarchies of the organizations, that the alleged activists subscribe to, are paid handsomely by those very corporations. Nothing dramatic is ever accomplished by these folks, so they glorify tiny, questionable accomplishments, instead, in their efforts to guilt other western democratic liberals into sending them a few bucks every month. Guilt, not charity, fuels western democratic liberalism. It's in the interest of these corporations for you to believe that these fake activists are heroes or 'rebels.'

Public Radio's and Public Television's invasive quarterly, week-long local pledge drives, are another case in point. They openly admit that the money they extract from 'subscribers', only accounts for five percent of their operating budgets. The other ninety-five percent comes directly from the foundations operated by the oil companies, drug companies, chemical companies, mining companies and other interlocking corporate subsets, that are actively involved in raping the planet and underwriting wars, strife, the destruction of families, the prison and mental hospital gulags, genocide and the expansion of National-Socialist bureaucracies.

Since PJ folks, especially these ex-hippies, essentially live in demographically pre-packaged dream-worlds they aren't troubled by these glaring contradictions, of course, and when you draw their attention to some of these contradictions, they launch themselves at you with all their claws extended and assault your character.

Fake activism is just another aspect of cultural schizophrenia and mass spiritual slumber. I won't be surprised if environmentalists, en masse, will be induced to make a move against the gifting network, nor should you be surprised if it happens, because we're quietly fixing all of the 'grievous problems', which they can only whine about.

Their attacks on this network would only be more free advertising for us, because the younger generations already look askance at their parents' and grandparents' silly, nineteen-sixties anarchy-posturing tales and their present-day, ossified nazification. The new and expanding demographic, among gifters, are the folks who are in their late teens and early twenties. I wondered when this would happen, because most of us in the vanguard, until last year, are middle-aged. The gifting process itself is essentially youthful, and it's rejuvenating for anyone who will do it.

Chapter Six

'She Blinded Me With Science'

Who developed orgonite? I'd like to know. Karl Welz was making and selling it by 1998, in a radionics application. I found out about that, and then applied the simple material to some new tasks.

Stacie Almquist told me that she met an old black man at the beach in North Carolina, who told her that the reason he was catching plenty of fish in the surf that day, and the other fishermen weren't, is because he uses sinkers made of 'magic rock.' He explained that his grandfather taught him to make it by combining warmed, softened pine pitch, tiny bits of cut-up can and an 'energy stone,' (rough quartz). The material is malleable, when warm, and he shapes it around the fishing line. That implies to me that orgonite was around during Reich's lifetime.

My sister-in-law, Melody, makes orgonite with pine pitch, collected from her property's forest in Idaho. The devices are quite powerful, according to psychics, and have an added subtle energy element. It's Melody who introduced me to Carol, in the summer of 1997.

Carol and I personally found out that fish are powerfully attracted to orgonite. Anyone can track this with a fish finder, but we first noticed that last summer, after we gifted a polluted, landscaped fountain in Spokane, Washington.

Since the structure was so beautiful, and someone had obviously put his heart into it, we wanted to clean the water, so we tossed a towerbuster in the middle of the lower pond, under the waterfall. A week later the water suddenly cleared, and we saw a couple of large koi, swimming ecstatically around the towerbuster. Koi, (big, fancy goldfish), are normally more lethargic. This is an experiment anyone can perform. We've cleared lots of bodies of water with orgonite, but that was the first time we noticed the effect on fish.

That day, there was a small crowd of people staring into the water, including some of the family of the fellow who made the fountain. They were awestruck, actually, and when I told them how we did that, one of them said, 'He's tried everything to clean that water and it only got worse!' This is the awe we usually feel, ourselves, when we perform atmospheric healing and see the effects, so it was nice to see 'regular' people having that response for something a lot smaller.

To get the 'fishfinder' confirmation, go gift an area by boat, where you aren't seeing any fish in your fishfinder, (sonar scope), and then go back there a week later and witness the new profusion of fish. In addition to seeing plenty of fish, you're also going to find that the water is considerably clearer and cleaner.

Orgone is the basis for all matter, energy, and life. It exists in various stages of refinement. It can be seen by energy sensitives as every color of the spectrum, the baser being red, (the force of decay), and the finer being violet, (the force of awareness). The energy of orgonite is mostly blue, (life force; water ether), as I mentioned.

One of the rare times I've clearly seen energy, was for several weeks after I made my own Big Secret, in January 2002. That device is the simple crystal/coil form, that Carol says she got from the little Andromedan, the night we were passing by China Lake, after leaving the first of two orgonite cloudbusters and several holy handgrenades in Death Valley, which looked completely dead at the time.

I saw a spaceship as clearly as the moon, for about an hour, as it also moved south in a parallel course, on the far side of the big, long valley, where that notorious underground base is. Carol was trying to sleep in the truck cab, and was irritably pushing the little guy, (who was invisible to me), away until she relented, and listened to his instructions, which she relayed to me. It was after midnight and we were successfully eluding the NSA trackers, for a bit.

I made one for her after we got home, which was right before I left for my first trip to Africa. I made one for myself when I got back, and saw lovely, magenta-colored energy along the entire length of the coil. It's very simple to make and if you want the instructions, you can get them from the www.ethericwarriors.com tutorial section. Carol uses a small variation of that, in each of her dolphin balls, using Herkimer quartz crystals in that case. She tells me that the cetaceans and Andromedans are closely related, which is why the dolphins like the energy from the balls so much.

I hesitate to mention aliens, because my enemies scour my writings for these mentions, in order to use the comments out of context, publicly, to 'prove' that I'm a crackpot and not to be taken seriously. By now I hope you've done a little experimenting with orgonite, and have seen some of the profound, but subtle effects, because I'd like for you to take Carol and I seriously. Disregard the comments about aliens, if they make you uncomfortable—I'm not trying to sell you anything with these comments; I'm just stating it for the record and for the sake of people who have had similar experiences and were told that they're crazy, by the PJ folks.

When we got that sweet Zodiac, in December, the dealer advised us to break the motor in for twenty hours, before taking it out in rough water. I confess that we only used up about four hours gifting the inland waterways near our home, before leaping out to sea, from Palm Beach Inlet, for a ten mile run back to Jupiter, where we live. And a couple days later, we drove the boat into the ocean in the other direction, from St Lucie Inlet, south to Jupiter, both days being fairly rough. Right after we went out Palm Beach Inlet, we saw a Coast Guard orange Zodiac, about the same size as ours, moving south, about 40mph into the breaking waves. The crew, who might have been new swabbies getting a little initiation, looked positively stoic and a little miserable, in all that spray. We were going north to Jupiter Inlet, at about 25mph, with the wind and seas at our backs, and even that was like riding a bronco.

This bronco-riding treatment is actually pretty good for a person who has a bad back, though you might not think so. I had back pain a lot, before I got my first zapper, ten years ago, and the two-day, violent ride through the hurricane in my little dory, before that, actually made it better. JFK, who had back trouble all his life, discovered this peculiar therapy, when he skippered a PT boat in the Pacific. He suffered a lot of back pain throughout his life. Lots of my customers have told me that the zapper stopped their back pain, but I don't have a clue how that works.

In December, we visited Dennis Griffin, in Charleston. Jeff, Dennis, Carol and I gifted Cooper River, a source of pollution and radioactive waste, and all around Isle of Palms and adjacent Sullivan Island. We tossed several hundred orgonite devices in the sea, river and intracoastal waterway.

Two weeks earlier, before the boat was delivered, Carol and I had visited Dennis, and he took us to the bridge, over the narrow inlet that passes between the two islands. A pod of dolphins greeted us there, and clearly told Carol that they wanted us to clean up that water, which is why we went back there so soon. This pod of bottlenose dolphins has dark, dorsal fins and lighter gray bodies. They also told her that they've known Dennis since he was a boy, and would really like for him to get in the water with them.

One of those dolphins showed up at the mouth of Cooper River, in Charleston Harbor, (on our second visit), when we'd just finished gifting five miles or so of the river along its length, and the rest of them showed up in that inlet under the bridge the next afternoon, after we'd completed the circuit. They were moving a lot faster than when we'd seen them, a couple weeks earlier and also came closer to us. Carol said they wanted us to know that we'd done a good job, and that they were grateful to us all. Dennis and Jacques Juarer, who since moved away, had changed Charleston into a very pleasant place, with their very thorough gifting campaigns on land.

When you toss more than a few gifts into a body of water, it will probably be a week or so before you see the confirmation of cleaner, scum-free water, but you'll likely get a visit in the sky by sylphs, immediately, which is always a nice confirmation, too. For some reason, they seem to particularly love it, when we gift inland bodies of water. We don't see them as often over the ocean, at least not yet.

Thousands of people around the world are discovering this water-clearing phenomenon, through their own experiments now. The best photos of Sylph cloud-forms, that I'm aware of, are on www.ryanmcginty.com. Ryan's a very observant and skilled photographer, and he's also energy sensitive. If you notice the difference between these and the fake sylphs photographs put on the internet by disinformants, you'll be on track.

The fakes are good secondary confirmations, like the weak reprisals by government predators are, but I'm not going to promote the fakers here, by mentioning their names. The most insidious one has recently abandoned his assumed 'Sylph spokesman' role by now, (he's finally lost credibility), and is being promoted on the web as an 'orgonite expert,' instead.

Some believe that Sylphs are elementals, and others, including Carol, believe that they're simply a specie of non-physical entities, though she calls them, 'angels.' We act on the assumption that nothing in our physical world can happen without the active involvement and endorsement of elementals, all of whom adore orgonite. They apparently also adore all of us for tossing it to them. I've seen their shimmery forms a few times, so I believe in them, even though I'm not particularly psychic. If you want that kind of certainty, you're going to have to earn it by giving them plenty of orgonite. There are no free rides or shortcuts in this network, though the entire effort is essentially a shortcut to empowerment and increased accountability. We're too close to it to see the scope of this miracle, and we're all struggling to dismantle a lifetime of self-limiting programming.

Many reputable people have seen orgonite devices disappear in clear water, right after being tossed, and the psychics tell us that undines (primary ocean elementals), personally come to take the orgonite away to appropriate spots in the sea, as soon as it's submerged. They've also seen the undines rise up in front of them and thank them.

Maybe you feel like you're drowning in questionable information by now, and that I've just tossed you an anchor. I certainly commiserate with you, if so, but remember that nothing I say is authoritative. I'm just recounting our understanding of reality.

As we see it, the elementals are neither good nor evil; they're simply responsible for keeping the physical universe running.

My encounter with mischievous sea sprites, during my most recent travail at sea, five years ago, confirmed that for me; I was miserable, and I heard laughter, right before each wave crashed over me during a storm, after my boat had foundered on the Grand Bahama Bank. For two days and nights, as I slowly drifted south, toward an uncertain fate, (Cuba and imprisonment?)

Muhammad is the only Prophet Who specifically commented on elementals, as far as I know, and He called them, 'Jinn.'

'Genie' is 'Jinni' in Arabic: 'of the Jinn.'

He calls them 'people made of clear fire, who existed before man was created.' The clear heat shimmers that I've seen, when Carol and Nora pointed elementals out to me, are in line with that. When Jeff psychically asked for the name of the primary elemental, which attached to his interactive, horn-shaped device, he got 'El Jeno.' It's pronounced 'El Haynyo', but I don't have a key for the umlaut that goes over the 'n.'

That is Spanish for 'The Genie', and Jeff clearly sensed that this enormous undine is near Yucatan, in the region where HAARP boosts their HAARPicanes, as they move into the Gulf of Mexico from the Caribbean. He had to look it up, because he's not familiar with Spanish. At first he thought he was hearing something close to 'Henry,' so we called it that for a while. Elementals don't have a physical form or human names, of course; all of that is a convenience, or perhaps a courtesy, based on the observer's cultural orientation. Carol favors the classical winged-fairy image, for instance, but she clearly saw menehunas, in their traditional form in Hawaii, before someone told her about them. They showed up when she was gifting around a big underground base, on the Big Island a year ago, then again a few other times. She did the underground base in the mountains, because the coastal area felt and looked very degraded, when she arrived, and even the dolphins were lethargic and disinterested. After that, the area looked and felt like Hawaii is supposed to, and she had some profound and lively interactions with the dolphins in the bay.

I think it's a bounty to be able to see the cloud-forms created by the sylphs, because seeing them, which triggers the heart connection to these creatures, doesn't require any psi ability. According to my Oxford Dictionary, a sylph is 'an imaginary spirit of the air,' and that the Germanic root word indicates 'forest.'

'Undine' is apparently a term for 'female water spirit,' invented in the 16th century by Paracelsus, based on the Latin root word for 'wave.' Maybe Paracelsus was psychic and saw them. The undines, which the psychics are seeing, are huge, though, and they haven't mentioned gender. I don't imagine that gender is much of an issue, for non-physical entities. Sometimes a Sylph cloud sculpture fills half the sky; sometimes there are many smaller ones. I think we all need to adopt a 'wait and see' attitude, in expectation that we'll get more information about these things later on, because it's pointless to theorize much, unless we have enough information to have an opinion, or are aware enough to understand the answers we get.

If you're not tweaked beyond credibility by what I've been telling you, you might enjoy hanging with some reputable psychics, when they're around orgonite devices. They'll carry on a conversation you can't follow, because they're talking about subtle energy dynamics, which you can't see unless you're gifted, too. They're obviously seeing the same things, so that, too, can be a confirmation for you.

It used to be smoggy here, until Jeff disabled all of the HAARP and death transmitters, from Palm Beach to Cocoa

Beach, which is 130 miles north. When you figure that there's a HAARP array every twenty or thirty miles along all primary highways and coasts, also a new death tower for every 2,000 people in most countries, that's a lot of orgonite and legwork for any diligent gifter who has a social conscience.

I'm sure that if he hadn't done all that work before Carol and I got here, that Wilma would have been a genuine hurricane, when it arrived at 'Ground Zero,' which is our little town. Fortunately, Jeff's previous gifting and the cloudbusters he made and distributed locally, knocked it down to a category three storm, and no serious damage was done around here, then. The significant storm damage was done farther south from the center of the storm, in Ft Lauderdale, as I mentioned before, and damage decreased north of the center, which was our little town. We'll get to the death/HAARP forest of towers to the south, also the HAARPicane Control Center, an underground military facility in Miami, which the What To Think Network calls, 'The Hurricane Tracking Center.' We've prevented volcanic eruptions and earthquakes, with orgonite, so that HAARP cancer under Miami shouldn't present a problem for us.

In early February, Jeff, Eric Carlson and I covered the rural, accessible areas of entire peninsula, south of Lake Okeechobee, on a coordinated gifting run, though we'll need to do the coastal cities, in order to finish the job. Carol and I will cover the coast, along the Everglades National Part soon, too, in our boat because there are no roads in that region. She's certain that there are major HAARP facilities in that area, which need to be disabled.

Dave Emmett, in Barbados, does so much business in Miami, that he bought a car and keeps it there—a real nifty Porsche Carrera. He's not confident with maps, so he bought a fancy GPS navigation panel, to find his way around the streets. It's been problematic for him, because the GPS shuts down, when it gets close to a military installation. There aren't many military installations listed in South Florida, but the proliferation of underground bases makes Dave's navigation device almost useless, because they're all considered military installations, and they're literally everywhere around here. 'Whose military?' one might ask.

Titan Corporation operates a facility south of here, out of which trainloads of sand constantly stream north, through our town. The north/south train tracks are right behind Jeff's condo. The trains have been coming out of Titan for a couple of years, but the hole at the Titan facility doesn't seem to grow, so it's all coming from underground. There's apparently even a train tunnel, that runs diagonally under those tracks, not far from Jeff's house, because he hears trains in there, too, and the ground shakes just like when a train is moving past on the surface.

I should probably tell Richard Sauter about this, because he writes terrific, well-documented books about underground bases. The occult/corporate 'military' folks used to just dump the dirt from their underground facilities near the sites, but I think they stopped doing that when Richard published his first book, in the early nineties. They still drop a lot of that to create 'garbage ziggurats'. These tall, multi-acre mounds, seen all over Florida, are half dirt and half garbage, after all, and the land is flat, so there's no genuine excavation taking place for highways or buildings. It's more likely that a building site will receive dirt, than give it away, since building sites need to be built up, in order to get above the flood plain and swamp.

Homestead Air Force Base, whose underground facility was apparently completed before the base was allegedly closed, after a big HAARPicane in the early nineties, boasts an artificial mountain, (ziggurat), over several acres of the still-restricted area of the facility. It was right after gifting that base, including spud-gunning towerbusters into a couple of huge, hard to reach nuke-cooling ponds nearby, that Carol and I were assaulted by a fake cyclone, or down-on-deck jet stream blast, in South Miami. We were passing a huge HAARP facility at the time, and the wind started suddenly and violently, right after we saw several flashes of green lighting, at that tower array site. Carol said that the Homeland Security Abomination was pretty mad at us, for gifting Homestead and were expressing their disapproval to us.

All traffic stopped on that little stretch of interstate for a few minutes, until the cyclonic wind and driven, torrential rain suddenly stopped. The local What To Think Network meteorologists commented on it, but didn't offer an explanation. I think they brought the jet stream down to the ground for a moment, which is something HAARP is capable of in very small, isolated incidents. I bet it costs them dearly to do that, in terms of risking exposure. The only other times I heard about that, was ten years ago in Exeter, New Hampshire, (blew most of the trees down in a small area), and again near Oroville, Washington in September 2001, right after several orgonite cloudbusters were deployed there. At the time, there were very few cloudbusters and there's a massive HAARP facility just over the border from Oroville, in Canada. There are a lot of particularly heinous HAARP facilities along the US/Canadian border, on the Canada side. I think the Canadians bend over a little farther and more enthusiastically, for National-Socialism, than Americans generally do, because they get more alleged material benefits, including easy, free access to serial killers, (MDs), and welfare for anyone, as an easy alternative to heavily taxed employment.

Usually, when a person of conscience finds out, from experience and observation, what orgonite can do, he or she commits to gifting entire cities and even regions, states or small countries. Young Satish Binda, who has been gifting Amsterdam extensively with Dr Dirk Verelst, is in the process of buying a car, so he can do all of The Netherlands, only some of which has been covered by other gifters, by now.

There are so many doing this and not reporting their accomplishments, that it's probably a good idea to determine which towers haven't yet been disabled. That's not a problem for a psychic or energy sensitive, but those of us who can't see or sense energy accurately need an edge. Carol taught Jeff to see the energy of a busted tower, because he's got a talent for seeing subtle energy. In that case, psychics see a dark outline a few feet away from the tower, when the transmitter has been gifted. The energy field beyond that dark line is bright and vibrant. More people can see deadly energy, than can see healthy energy, but seeing the dark stuff is certainly sufficient for this purpose.

When I went on my first large-scale gifting expedition during August 2002, I specifically went alone, because I wanted to see if the confirmations would be consistent and obvious enough for non-psychics to see. I was lucky, in a way, because nobody had been busting towers yet in Southern Idaho, east of Boise. Jerry Morten and his son, Ben, who has invented terrific energy devices with orgonite, had already disabled most of the new and old death transmitters and HAARP arrays, in and around Boise and later finished off the rest of that part of the state, to the west and into Southern Oregon.

Southeastern Idaho, a huge area that's mostly made up of the wide Snake River Valley, was a good test bed, because most of the population, therefore most of the towers, were along the Snake River, and the towns were uniformly spread along that valley. They were small enough that I could do one town and surrounding area per day, and at the end of each day there was a requisite thunderstorm, just over the town. The region had been in a severe, HAARP-induced drought, for quite a while, so the rainstorms were that much more confirming.

Also, this area was beyond the range of Jerry's cloudbuster, and nobody to the east had one yet, either, as far as I knew. Yellowstone National Park and sparsely populated Western Wyoming, borders the east of that region, and at the time, the next-nearest orgonite cloudbuster was apparently in Salt Lake City, hundreds of miles to the south. It's important to have a few controls in place, when we experiment, otherwise the variables can make our data unusable. Cloudbusters network with each other and create some synergy, if they're closer than a few hundred miles together.

That's when I developed a reliable dowsing technique, because sometimes the locations of the towers, which I could always see from a distance, were so remote that only dowsing got me onto the right roads. Before, I was crummy at dowsing, but necessity is a good teacher. Here's the technique I stumbled on:

Put the point of a Succor Punch in the palm of your non-dominant hand and use a pendulum with your dominant hand. If you're already familiar with pendulum dowsing, you'll get proficient fast; if you're not, get a little book on dowsing from a new-age bookstore and a pendulum and familiarize yourself. It's not difficult or weird. You can make a Succor Punch from the tutorial on www.ethericwarriors.com or buy one from one of the vendors listed there.

I asked Carol to watch the energy dynamics, while I was using this method, and compare that with my unassisted dowsing. When I used the Succor Punch, she saw clean energy/information moving from the Succor Punch crystal, into my palm, over my shoulders to the hand holding the pendulum.

When I wasn't using the SP, she saw clean energy/information coming down, into the top of my head, but it got altered and a little polluted, by the time it came out my neck and over to my hand. She says that this is because whenever we try to express something that we receive, our brain usually encodes it in a 'comfortable' way for us, and that the brain is always prejudiced. We temporarily overcome that filter, when we're particularly lucid or detached, (or in dire need), but dowsing should never be considered infallible, any more than a single psychic's observation is. It's an aid.

When you're out on a gifting sortie, you're selflessly motivated and can be functionally detached, without much effort or angst. This is a good time to test your Succor Punch aided dowsing technique, I think, to determine if someone has already gifted a certain tower. I call this method, 'turbodowsing.'

Human parasites are similar to parasites in the body in many ways, including this one: they get sustenance from deadly orgone radiation. This energy both causes and is produced by decay and human misery, and the process is self-perpetuating more than health is, which is one reason we're constantly in need of healing. All things in the physical world naturally decay, after all. The way internal parasites create this dynamic, is by excreting acidic toxins into their environment, which causes a net positive charge; a deficit of negative ions. The underlying energy matrix of a positive-charged environment is deadly orgone radiation.

Healthy tissue is slightly alkaline, which is characterized by an abundance of negative ions, (electrical charge). A stasis is reached, and when more ions come into the environment after that, they're simply moved along. This is why it's impossible to harm oneself with properly made colloidal silver or with a zapper. Deadly orgone radiation, steadily decreases the number of ions until there are none, which means that parasites are ultimately suicidal. A big fluke in a corpse's liver won't find another host, for instance, after its host has died of hepatitis or cirrhosis.

Parasitic people instinctively and willfully make others unhappy, whenever possible, because they feed on others' misery. The world order is parasitic, so is mainly staffed by human parasites. I use the term, 'human,' advisedly, but to my mind if someone looks human, he is human on a practical level. To believe that any specie is essentially evil is to buy into the old dualistic programming that has kept humanity's neck under patriarchy's ideological boot for millennia. Devil worship, which can be called infantilism, is the ugly cousin of dualism, the false but persistent, 'Sky God,' doctrine that leads people to assume that 'physical existence is evil, so we mustn't concern ourselves with whatever happens in our world.' If you want to see that schizoid ideology in action, visit any gathering of new-agers.

We're rather finding out, though some of the psychics, including Carol, have always known this, that all of creation is loaded with buried keys and treasures, that can be exploited for our own spiritual and intellectual progress, while we're here. Think of the physical universe as a university for our training, especially since all physical phenomena simply mirror more subtle, but significant, etheric, (spiritual) phenomena. Muhammad said, about man, 'Dost thou reckon thyself a puny form, whilst within thee the universe is enfolded?' Theoretical physicists claim that the universe is holographic and that the sum is contained in all of the individual parts. Jesus said, 'In my Father's house are many mansions.'

I don't use the word, 'evil' much; only ever in the context of behavior. For instance, the world order is a corrupted and corrupting institution, founded on evil behavior and thrives by exploiting humanity and our planet's physical and etheric resources. If we can detach from a deep-felt need to judge the characters of human parasites, we also recognize their peculiar DOR-dining characteristic in nature itself, though natural creatures' intentions are never to exploit; only to survive. There's a lot more symbiotic, (mutually assisting), behavior in 'the jungle', than there is predatory behavior, contrary to The What To Think Network's wishes. The occult/corporate world order can be compared to vultures, for instance, which are attracted in large numbers, to the energy of decay.

Fish, like some cats, can actually transmute DOR, apparently, and are attracted to it. We figured that out when Carol's dad told us that he always catches more salmon, where high-tension power lines cross the Clearwater River, in Northern Idaho. It's obvious that those salmon are charging their batteries, resting in that DOR field, on their way upstream, from the distant Pacific to spawn. Some cats rest on top of televisions that are running. Some of us have seen the bad guys' flying saucers traveling along high-tension power lines and hovering over huge death transmitters.

A friend of Jeff's, who has a large, chartered research vessel with expensive, stereoscopic sonar, told him that he's lately seen huge towers on the bottom of the ocean, in the deep water around South Florida. I think airplanes and other capital will be needed to disable this underwater network, which is perhaps worldwide, but that will no doubt happen after orgonite reaches mainstream awareness. I don't know if it's feasible to drop enough orgonite into the oceans from the shore, to get this done. The network is handling the death towers on land, because we're so spread out, but it's possible that there are many times more of these transmitters on the seabed, evenly distributed, and most of those areas are outside of shipping lanes or air travel corridors. The old, parasitic world order—the source of all the new death transmitters on land and in the sea—is already exposed, but defeating and bringing them to account is going to take some time and patience. Thankfully, disabling these predators is fun and rewarding work, and relatively risk-free.

Another reason I'm writing this book is that I have a feeling that orgonite will become a household word, sooner than later and many hearts, (including Carol's and mine), might be broken, if the phenomenal success of this grassroots network of gifters were to be eclipsed by others' corporate, slick marketing and by coordinated, steamroller disinformation efforts, by the What To Think Network.

We've perhaps just gotten a glimpse at how this might be done, in fact. The Coast to Coast website, which we consider a CIA disinformation asset, is offering for sale, a pretty good orgonite device that's being advertised as a 'wine enhancer', and the ad makes no mention of orgone, orgonite, Dr Reich, or Croft. Carol's unique, flat spiral coil form, which she adapted, five years ago, from a crop circle design that showed up the previous day in England in June 2001, is even featured in the device. This has come to be called, the Saint Buster's Button coil, because it's normally used, along with a rare earth magnet, in a simple orgonite healing device, which Carol invented then, and to which I later gave that name. She uses a three-dimensional version of that coil in her Crystal Harmonizer, which is her latest version of the device. That's all shown on our site, www.worldwithoutparasites.com.

Another slick-looking disinformation website features a fellow, who calls himself a journalist and is in the business of slandering several pioneers, including me, though we've never met or corresponded. Another sabotage attempt was the fake, 'Don Croft style' zapper that was being promoted last summer, on E-Bay, and got a lot of attention until we debunked it. The upside of seeing these attacks, is that it gives us a better idea of where our potential weaknesses are.

The other side never comes directly at you, when they want to destroy you. They come from the side and from the back. This is how all parasites and scavengers operate, and we should never take their assaults personally.

More recently, an orgonite object was described on CNN as a 'terrorist device,' though they made a point to say that it's harmless. I suppose a toothbrush is a terrorist device, if a terrorist brushes his teeth with it. The Airport Gestapo seizes nail clippers from passengers, and I wonder if that's because a terrorist was once caught in the act of trimming his nails.

The internet contains thousands and thousands of pages of personal accounts, from reputable gifters, all of which make up an impressive body of empirical evidence of the transformative power of orgonite. Most of those are archived, only, unfortunately. The previous internet forums were run by saboteurs, and they erased all the substantive stuff, before they finally shut down the sites, due to lack of traffic. Maybe some enterprising computer mavens will soon exhume all of the disappeared reputable reports from archived HTML files and put them back into play. If you want to do that, we'll be happy to host that material on Etheric Warriors. It would also be nice to get some of the French and German reports translated into English. I think those two boards will remain viable, along with ours.

D Bradley produced "CHEMTRAILS: CLOUDS OF DEATH", four years ago, which is not only a singular, lucid film documentary of that failed bio-war agenda, but also tells how to disable the chemtrails where one lives, by building an orgonite cloudbuster, which thousands of people in North America and Europe had already done, by the time the film was released on July 3, 2002. These two continents sustained the bulk of the chemtrail assaults.

I think that a fairly popular book will ensure that the network which Carol and I initiated, almost five years ago, will keep its well-earned place in history, rather than becoming an overwhelmed, marginalized footnote, due to massive disinformation and slander campaigns, and that the many bright lights within it will be enabled to inspire and educate millions of awakening people, in coming days, rather than be overrun by the What To Think Network.

I think we're going to learn, pretty soon, whether some of the orgonite we've been tossing into the sea, locally, is being taken by dolphins to gift the larger, very deep transmitter arrays in the Gulf Stream that show up on the fancy, sonar devices. Carol has a strong sense that this has already happened, but she wants us to take a trip across the Gulf Stream to the Bahamas, in order to get a closer look. We dowsed the location of some of the major underwater transmitters between South Florida and the Bahamas and when we finally get out there, Carol will be able to determine if the sites have been disabled. That's hard to track directly, when the sites are beyond the horizon.

The discovery of one of my old friends, five years ago, that water amplifies the effects of orgonite, came later than the occult/corporate discovery that water also amplifies their own deadly energy. Carol feels that the new challenge for us is to disable enough of these ocean death transmitters, to allow the dolphins and whales to remain in our seas; otherwise they'll perhaps just leave our reality. Wouldn't you leave the area, if someone filled your home and neighborhood with poison gas?

Judging by how eagerly they want orgonite devices, we feel confident that we have a chance to succeed. Imagine the historic precedent of a 3D partnership between our species, in which any of us on land can participate!

Carol said that during her last swim in Hawaii, this time, one of the spinner dolphins gently 'bumped' her foot, repeatedly touched her body as he swam alongside, turned around to look at her energy for a moment when he had swam ahead of her, then swam away. By the time he turned to look at her, all of her pain had suddenly gone. She'd been under considerable etheric assault for several days, because the bad guys, as usual, didn't want her to get into the water with those dolphins.

When she arrived, a few days before that, Nature's Conservancy were wrapping up their dogged agenda in that bay, and the ambient energy from their effort was dreadful there. I think someone had tipped that organization off that many people spontaneously swim with dolphins in that bay. All the dolphins had left, when the Nature's Conservancy boats arrived, and there were suddenly a lot of tiger sharks and stinging jellyfish in the bay, which was quite untypical. She had expected to just jump in the water and get busy with her cetacean friends, so she was pretty disappointed and a little disgusted by those self-righteous Nazis. It's said that, 'Like seeketh like, and taketh pleasure in the company of its kind.' I wonder if these thugs keep poisonous snakes, leeches, hyenas and scorpions as house pets.

On the second evening, after failing to connect with the dolphins again, she drove across the island, over the volcano, to visit Dr. Len Horowitz. At his request a year ago, during her previous visit, she had healed the poisonous energy matrix on his property, by finding the etheric trouble spots, and putting orgonite in each one. He had hired some 'native shamans' to clear the energy before. They were recommended by some of the Doc's new-age acquaintances but the 'shamans' only made it worse. Go figure. I don't think those guys are part of the Kahuna network, because Carol said the energy of the volcano is peaceful, which indicates that integrity prevails among the genuine Hawaiian Kahunas.

This time, he had asked Carol for a special orgonite device for the steam vent on his property, around which he built a steam bath. Carol made the device, with Ryan McGinty's generous help, and put it in the steam vent. Dr. Len had bought that phenomenal property from some Illuminists, (Satanists), not long ago and human sacrifice rituals had apparently been taking place at the deep steam vent, which is the resort property's centerpiece.

While at the top of the mountain pass, she left a similar, specially made orgonite gift for Pele, the volcano's guardian. The sun had set and it was dark, but she was urged to leave several other orgonite devices there, too, which puzzled her. She complied, of course.

On the way back, in daylight, she saw a big HAARP array and a weather-ball near that spot, which the orgonite had disabled the night before. The native Hawaiians have lately kept the energy of that volcano in good shape, as I mentioned, but more good energy is always helpful, and Carol told me that Pele had asked her to do it. This was the first time she ever made red orgonite. The constant, HAARP-generated rainfall, throughout the region, stopped after she disabled that big HAARP facility in the mountain pass.

The next time Carol got out into the bay from the beach, the dolphins had returned and most of the sharks had left. The Nature's Conservancy people were still out in force, though, warning swimmers to stay away from the dolphins or risk being 'bitten by wild dolphins.' Apparently, environmentalist Nazis exude a sort of etheric B.O. that has a social effect, like a noxious fart in a crowded, etheric elevator. We heard one of the staff, at the dolphin facility in the Keys, warning the crowd that 'wild dolphins are dangerous,' too, so maybe this is the new party line for institutionalized, fake nature lovers. It's a lie, of course; nobody has ever been harmed by a dolphin or whale.

Carol feels that they drove the dolphins from the bay, with their collective unpleasantness, and the dolphins confirmed that for her when they returned. The environmentalists also attracted the tiger sharks and stinging jellyfish—nature abhors a vacuum, after all.

In my view, stopping the Rockefeller and Oil Cabal money that liberally funds Nature's Conservancy would do more to 'protect the wild dolphins', than their benighted efforts ever could. I hope that this book will help expose their real agenda, and help neutralize these fanatics' malicious efforts to prevent humans and dolphins from connecting on the dolphins' turf.

> Fanatic: one who, having lost sight of his goal, redoubles his efforts to reach it.

Chapter Seven

Definitive Signs and Tokens

It seems weird to me, how you can see dolphins on the surface once, then might look all around for a half hour, and fail to see them come up for breath again. Carol says it's because they do what the Sasquatch do and that they do it often—simply sidestep our dimension or timeframe. I think the solidity of our existence is called into question by this practice; also it seems easier to relate to how ghosts and other supernormal entities can be here with us, but not seen with eyeballs.

That might not be as mystifying as you would suspect. Many of us in the network have encountered federal agents, who are able to functionally disappear momentarily, apparently with the help of a hand-held electronic device. We gather that they do this so that they won't be photographed. The most dramatic case I experienced, was when four of us were on a sortie, driving earthpipes into the ground, around Ft Lewis, the big army base near Tacoma, Washington. Ryan McGinty got out of the car to drive in the last one, and a red jeep pulled up and parked on the other side of the otherwise-deserted back road. The rest of us watched the vehicle closely, and tossed energy at whoever was inside. Red fedmobiles are most often driven by secret police psychics, and we were doing a lot of damage to Ft Lewis' underground facilities that day. Ryan got back in the car, and we drove slowly past the Jeep—nobody was inside. The doors never opened, so we're pretty sure nobody saw where Ryan did the deed. We all had a sense that, whoever was invisible, was knocked unconscious by the blasting. Psychic predators are particularly vulnerable to energy that's tossed through the heart.

I once knocked out an FBI guy, who was tailing me on a

Greyhound bus, a couple of years ago. He and his partner were sitting a couple rows away from me, on the other side of the aisle. Carol and I have been under box surveillance, for the better part of five years, and we get annoyed by that, sometimes. I'd made the guy and simply threw energy at him, then he slumped over, unconscious. He was still breathing, but nobody could wake him. The driver stopped the bus at the next highway exit, and he was carried away in an ambulance. His partner looked pretty intimidated, but he wasn't that susceptible. He got off at the next stop, and gave me a peculiar look on the way out. I don't think the sewer rats told these guys anything about us; otherwise they'd have sat farther away from me.

We all disappear from their electronic and satellite surveillance, when we turn on a Succor Punch, and that probably still mystifies them; it certainly annoys them, because having to follow us with their eyes and in several vehicles, means we can see them, too, and can then throw energy at them. It's fun to watch them race to highway exits, after we make/blast them. Try it! There are apparently millions of plainclothes Homeland Security Abominations, (probably mostly Russians and Eastern Europeans by now, considering who's running that American Gestapo agency), that are actively surveilling countless thousands of American citizens. It might be a time of plenty, for the dry cleaning industry, as more and more of us little guys learn to toss energy at these traitors and invaders, and make them soil their suits.

When we weren't watching, the Bush Senior regime drastically reduced the number of potentially patriotic military people, in the early 1990s, while exponentially building up the secret police agencies. This sort of sleight of hand characterizes American history, since the European bankers formally arranged to take over the US Government in 1910, at Jekyll Island, Georgia. That area is heavily gifted now.

The most highway-spook fun we've had, so far, was when a cloudbuster was on the Jeep's roof rack, on our latest cross-continent driving trip. The spooks behind us, in that case, just about wrecked, trying to get in front, when the CB pointed toward the back; when it pointed to the front, the box-surveillance team members up there drove erratically and exited, at the first opportunity. We've literally raced them to the toilets at rest areas, just for a hoot. 'The Keystone Homeland Security Abominations.' I should get this on film one of these days.

The first time we used an orgonite cloudbuster for political reasons, was during the 'Show Us the Law!' demonstration, outside the headquarters of the Internal Revenue Service, (a 'for profit' corporation in Washington, DC), in early April 2001. Irwin Schiff, who was lately thrown unceremoniously into a federal prison for nothing, led that demonstration from his wheelchair. He graciously accepted one of our Terminators that day. The Director wouldn't let Irwin into the building, when he knocked on the door. The schmuck just stood inside the glass doorway, with a high-level Washington cop, and smirked at Irwin, who is an awfully nice fellow and a thorough research journalist.

In that case, Carol and I arrived at 3AM to get a good parking space, close to the IRS director's third-storey office in the corner of the building. We aimed the cloudbuster through the front windshield at his office all day, and it was a wonderfully peaceful demonstration, even though some plants were trying to incite trouble among the crowd.

Even the cops had a good day, except the scores of black-clad SWAT cops who had to spend the day nearby in a cramped, disguised tractor-trailer. I saw into that trailer, when I was out walking around the Mall area at six AM, and they looked like they were expecting a war. I wonder what would have happened, if orgonite hadn't been present during that demonstration. Orgonite is powerful stuff.

When we arrived, the sky was overcast. Drizzle, wind and cold was forecast for the entire day, but by the time the sun came up, the sky was clear blue and remained that way. There wasn't a breath of wind all day. Our first cloudbuster was only a few weeks old, but we'd been getting that consistent result every time we set it up during our trip. We were on our way back to Florida from New England, that day.

A federal cop with expensive shoes and dark sunglasses stood near our car for quite awhile, monitoring the huge crowd of demonstrators from there, but he didn't seem to notice the cloudbuster. Go figure.

The finer, trans-dimensional ability of dolphins might explain why that pod of rough-tooth dolphins 'showed up', for part of a day in Haifa Bay, Israel, a year ago. It was apparently to alert our psychics, in a timely way, via the What To Think Network, (strange to tell, who were reporting the bizarre event), that there was a huge nuke bomb on the seabed in the bay, about to be detonated to precipitate World War Three. The psychics found an identical bomb on the seabed off the coast of Alexandria, Egypt, also set to detonate then. It got taken care of, too.

Rough-tooth dolphins are never seen far from the equator and they've never been seen near land before. Carol had gotten initiated by an enormous pod of these dolphins near Costa Rica, a few weeks before the Haifa event, which was poignant for me, because the Baha'i World Center is on Mt Carmel, overlooking Haifa Bay.

We, in the gifting network, are not the only people dolphins are showing their hyper-dimensional talents to. We're apparently just the only ones taking our cues and acting on them. If you read http://www.earthtrust.org/delrings.html, you'll see an example of dolphins attempting to educate humans about hyper-dimensional physics; in this case, 'educated' humans who apparently miss their cues and are even afraid of magic. Most people, who get big bucks for studying dolphins, are serving a pretty heinous agenda, because the sewer rats are paying these folks to help them 'know the enemy.' Dolphins have made it clear to some of us, that they don't like tyranny and exploitation, and they often participate on our predator safaris, on Etheric Warriors. Lots of cetaceans have been murdered by the US Navy and British MI6, on account of our new partnership with them, unfortunately.

As you can see by reading that article, materialistic scientists, just like irrationalist Theosophists, are adept at explaining everything without answering anything.

It's very rare that a scientist or scholar pays attention to who is signing his or her paycheck. I think that social conscience is as rare among scientists and institutional scholars, as it is among US soldiers, who refuse to honor their oaths to defend the US Constitution against domestic enemies, such as their Commander in Chief.

A few weeks before the rough-tooth dolphins showed up in Haifa Bay, Joe Vialls, a reputable research journalist in Australia, achieved quite a coup, when he posted some priceless intel about the bomb that caused the deadly tsunami in Southeast Asia. The What To Think Network had been trying to persuade people that the 'natural' tsunami in Asia was just a precursor to many more around, the world in coming days. When that agenda failed, thanks largely to our predator safaris, during subsequent months, they simply stopped talking about 'eminent, worldwide tsunamis,' just like they stopped talking about 'a new ice age from the eruption of the Yellowstone caldera,' after Carol and I surrounded the caldera with twenty-six earthpipes in October 2004 and stopped the seismic activity there. Joe Vialls was murdered soon after he released that information, I'm sorry to say. I wish we'd seen that coming, so we could perhaps have helped him avoid getting rubbed out, as we routinely do for ourselves and many others.

Carol, Jeff and I are currently involved in preventing the 'plague of destructive hurricanes' that the What To Think Network had been promising will happen here, very soon. When we're all done, disabling the HAARP network in this region, the What To Think Network will no doubt stop mentioning that 'imminent natural calamity' in the Gulf Region. They abandoned the 'imminent invasion from space' scam a few years ago—did you notice?

Fortunately, we had stumbled onto a way to neutralize the underwater H-bombs, with the help of the cetaceans, right after that one went off in Southeast Asia. We took our cues from Joe Vialls, that time. The psychics, with the cetaceans' help, found and we all neutralized about 80 of these massive nuke bombs on the seabed, throughout the world in the following months. These massive, hydrogen bombs were apparently all put in place by American, Russian and British subs, near populated coastal areas throughout the planet and sometimes on fault-lines, too. Another, smaller H-bomb was found in Mt Aetna and neutralized.

After a couple of guys in Rome—Igor Cinti, of www.orgoneitalia.com, and John Leach—dropped a trail of orgonite devices, from the ferry between mainland Italy and Sardinia last year, a pod of orcas were spotted in that area almost immediately. Their appearance was covered by the national media and astonished everyone. I'm not aware that orcas are commonly seen in the Mediterranean, but it's certainly uncharacteristic to see them in those waters, at least.

Rough-tooth dolphins are the specie that greeted Carol en masse off the coast of Costa Rica in December 2004. I say, 'specie,' but in fact maybe all of them are the same specie, like pygmies in Congo are the same specie as Nordics in the Arctic Circle. The rough-tooth dolphins had never been known to approach humans before that, (or since, according to Roy, the Costa Rican marine biologist who had driven the boat that day—Carol stays in touch), and they've also never been known to even approach land, let alone such a heavily polluted, ruined bay as Haifa's.

Another phenomenon of Carol's Costa Rican dolphin encounter was the presence of an exceptionally large rough-tooth dolphin, in the middle of a line of hundreds of them, who were swimming in a synchronized way toward the idling boat. The distinguishing features of these dolphins are that their teeth each have serrated edges and there's no bulbous forehead as with bottlenose, spinner and some other dolphins. Also, these are equatorial dolphins.

Carol told us that the spinner dolphins we've been encountering, at the end of our gifting sorties, are another deep-water specie, that's not common here.

On Carol's latest visit to the spinner pod, in the little bay near Captain Cook, Hawaii, she was treated to an astounding geometrical swimming display, which also happened on the day she was healed by that kindly dolphin. A very young, uncharacteristically white spinner dolphin remained suspended, vertically, near the white sand bottom of the bay, near Carol in about fifty feet of exceptionally clear water. Several other dolphins swam along the bottom from several directions, toward the white one, and then moved upward, in a double helix pattern, around the white one, from the bottom up to the surface nearby. She's got some photos of that event, also of the dolphin who healed her.

Before the pod showed up in this bay, if you wanted to swim with dolphins out in the open, you needed to pay a company several hundred, dollars to get you out to sea in a boat, for an hour or two. I think this is the first time that dolphins have made themselves available for casual contact, from a public beach. Well, it's a rocky shore in this case but you can get in the water there, because it's on the leeward side of the Big Island. I think that if I lived there, I wouldn't care to go traveling any more.

I think we're properly motivated; that is, we don't do all this gifting with the expectation that we'll swim with the dolphins, around here. We do it because it's simply right and necessary. There can't be another acceptable motivation in this life, of course.

Shortly after the Haifa episode, a pod of rough-tooth dolphins were driven ashore, apparently by murderous US Navy sonar assaults, (a reprisal?), in the Florida Keys, and some were then butchered. Was that sour grapes, on the part of the CIA and their Navy? In some cases, etheric warfare spills over into physical violence against the cetaceans, unfortunately, and we all have to be prepared to accept that, if it happens to us, too. Freedom was never free, though pretty soon maybe tyranny will end, if you and I do our parts.

Two of our African associates, Dr Rushidie Kayiwa in Uganda and Mrs Atieno Odondi in Kenya, have experienced physical assaults on account of their participation in this network, and several of us, here in the USA, have come to moderate harm and escaped a few murder attempts, by federal agents in the path of doing this essential, environmental healing work. We're lucky that Dr von Peters is available to get the toxic metals out of all of us quickly, with his wonderful, gentle herbal chelation technique.

Dr Muhammad Siyyid, a venerable businessman and healer, who had lately become a populist politician and clergyman in Qum, Iran, was gradually poisoned by other clergymen, then murdered in a hospital, shortly after he began gifting in that city, though to be fair, it was more likely that he was murdered for being a populist, in that benighted country, rather than for gifting. We couldn't get a zapper to him in time to perhaps save him, from his assailants, unfortunately. One can't simply 'mail' something to Iran from the US. The poisoning began before he started making and distributing orgonite. His ancient forebears, the Imams (lineal descendants of Muhammad), were all murdered by the Turks except, perhaps, the twelfth and last one. The current, London-sponsored Persian regime is no better, of course. It's a testament to his integrity, that he was open to receive information and help from a Baha'i, because Muslim clergy in Iran are responsible for murdering countless thousands of my innocent co-religionists.

Several gifters in the US, over the years, were broadsided by cars, which had sped through red lights, while they were passing through busy intersections but, thankfully, none of them was seriously hurt, though most of the vehicles were totalled. The psychics tell us that the drivers of the errant vehicles have invariably been active Monarch assets, remotely induced to get into a trance state, by their CIA handlers. That hasn't happened to anyone in this network for over a year.

I think the network has simply expanded too much, for the sewer rat agencies to easily get away with such rough intimidation tactics, especially because we report these incidents on the internet, when they happen to us, or to other gifters. The sewer rats hate exposure. When you take their pictures and post them, it's better than shooting them in the head, because they get into big trouble at work after that. Eric Carlson particularly enjoys taking those snapshots, and posting them. They've mostly learned to make themselves scarce, when they see Eric coming. After some practice, it gets easy to spot pavement artists, because they're so studiously nondescript. Nobody else you're likely to encounter in public is that nondescript and personality-free.

I'm not seeking to prove anything to you, as I've mentioned. If you feel, as we all do, that intelligent action is needed, right now, to undermine global tyranny, this book will help you steel your resolve, and you can learn a few useful tricks from it. I was a sign painter, before I got into the healing trade, and I can tell you that many constructive trades and professions are nothing but a collection of useful tricks, if one already has the ability to think critically.

A nice Dutch gifter, whom I correspond with, paid me a high compliment in email. She's slightly psychic, and lately has had some strange, but interesting visual experiences. She felt emboldened to tell me about them, because she knew I'd be open-minded, and nobody she knew, (most of us only know PJ folks in our daily lives), was willing to hear her. She said, 'You're the weirdest person I know.' I told her that Carol has me beat on that count, and that she ought to tell Carol some of her experiences, too, because I'm just a cheerleader when it comes to psi stuff; Carol and other sober, reputable psychics actually see what others describe, and can comment more intelligently. I also asked her to post about it on the internet, because this would empower others who have experiences like this, but are afraid to tell about it. You'd be amazed by how consistent people's preternatural experiences are. I remain amazed.

I think the illegitimate boards are all defunct now, except the Yahoo cloudbuster forum, which seems to keep drawing ragged breath, in spite of being populated mainly by paid and programmed sociopaths. If you want a negative example, take a peek at that board.

Getting kicked out of the entire Yahoo Public Forums site May 2002, after I posted the initial report about how orgonite can disable the new death towers, was the impetus that eventually led me to start my own board, in August 2004. I had been that particular Yahoo board's administrator, actually, and along with getting cut off, our then-current commercial site (www.worldwithoutparasites.org, also hosted by Yahoo) was suddenly inaccessible, and we had to hastily set up www.worldwithoutparasites.com. Two successive, private US servers subsequently allowed NSA hackers and leg-breakers to interfere with our business site again, so we eventually moved it to the same Montreal server that hosts www.ethericwarriors.com and things have been fine ever since.

That's not all that was taken from me that day, four years ago: all of the thousands of email addresses that were on my Earthlink email account were erased, and all of my business-related email was also systematically deleted by hackers at Earthlink, which shares a gated entrance and fortified compound, with the NSA's West Coast Headquarters. After a couple weeks of that, I started an email account with a local server, who also eventually bent over for the NSA, and wouldn't allow me to use the secure Phoenix email software any more.

I found it prudent to do all of my email without using email software after that; I do it all on the web now. There are NO email software programs that aren't accessible, one way or another, by NSA hackers, if you're using an American internet provider, and if you let them in, they'll destroy your hard drive. Nearly everyone in this network, who hasn't taken basic precautions against government hackers, has sacrificed at least one hard drive, as tuition. I encourage everyone who contacts me online, to stop using email software, because personally contacting me is apparently a federal offense, under the paradigm of the Patriot Acts.

They don't bother with courts, when they feel like punishing us all for non-crimes, under the protocols of the 'new patriotism.'

Earthlink's headquarters are outside of Los Angeles. The high perimeter fence they share has razor tape on top of it, and there are fortified guard towers in the corners. Yahoo! belongs to AT&T, a well-known CIA front company. They both put the squeeze on me pretty good, right after I spilled the beans about towerbusting, but all of that ultimately boosted my reputation, as you can see.

Kam Wong in Queens, gifted the AT&T headquarters in Manhattan and sent us a photo of that building. It has an enormous delta antenna on top, that's actually a separate building. Delta antennas are two pyramids, bottom to bottom, and the technology was developed at Montauk, as an aid to exerting mind control on the surrounding populace, also for giving physical substance to thought-forms. It's very close to Gracie Mansion, where the mayor lives. You won't hear about any of that on the What To Think Network, of course.

Kam also gifted the predominant skyscraper in Hong Kong, where the dark master, Singh, keeps a penthouse when he's in town, monitoring the global heroin and banking networks.

If the Homeland Security Abomination, which the NSA, CIA and FBI now serve, was as powerful as the What To Think Network wants you to believe, don't you think they'd have been able to stop this network's expansion in the early days, or at least to succeed at killing Carol and I instead, of constantly bungling their persistent attempts? I reported all of that, until November 2005, in the 90 journal reports that you can get from www.ethericwarriors.com without charge. My subsequent reports have been in posts on the same site, and I'll keep posting there.

It may be that I demonstrate an attainable standard for what one can experience in the weirder world, without being particularly psychic. There's no shame in not having that gift, and it's important to consider that most of the 'gifted ones' work for the CIA, British MI6, Mossad, the KGB octopus and it's sleazy Balkan tentacles, the Vril society, the Chinese equivalents, etcetera, so having the gift doesn't automatically accrue any personal glory.

Most of these psychics won't even remember their predatory activity, because they've been successfully programmed as dissociative, (split), personalities in the unimaginably vast mind control programs based at Langley, Montauk, Tavistock, North Korea, China and God-knows-where-else.

During their conscious lives, most of these psychic predators are probably neurotic, grinning, vacuous new-age minor leagues cult gurus or sycophants, spewing saccharine 'love and light' platitudes at all comers, and incessantly climbing over each other in ragtag Theosophical and Rosicrucian hierarchies to seize center stage as, 'most enlightened, selfless and spiritual.'

This sort of takes the bloom off the rose, if you think that being psychic equates with having discernment or being spiritually mature, don't you think? Did my karma run over your dogma just now?

As a rule, the most messed up people you meet are the ones who are most adamant, in their claims that they have all the answers you need. When they have a string of letters behind their names they reach the top of the various institutional hierarchies, because the occult/corporate hierarchies favor lettered predators and parasites: the new nepotism, now that the biological family has died a slow death.

If you think I'm wrong, consider all the psychiatrists, psychologists, psychology students, Christian fundamentalists and new-age proselytes whom you've been subjected to. If you're lucky, like me, you'll have known some genuinely helpful professional psych folks, like Doc Dirk, who's gifting Amsterdam these days, and uses orgonite for therapy.

Another exception to the rule is James Hughes, a lightning shaman, (received knowledge and profound psychic ability in the instant of being struck by etheric lightning), who has personally introduced thousands of people, during the past twenty years, to synergistic, effective uses of crystals, color, sound and subtle energy for healing, enhancing awareness and problem solving.

Whenever anyone gets into proselyte mode around me, I have an urge to take out my pistol. If you don't feel similarly repulsed by influence peddlers it means you've swallowed their endorphin bait, on account of your own residual mind control programming. They won't be ignored, because they're predators and the best thing to do when their foot is in your door, is to just shut the door. No matter how nice you are to them they're still going to try to scam you, because that's who they've chosen to be.

I can tell you that if Carol had been one of those cowardly, duplicitous, vindictive love-and-light Nazis who avoid personal commitment and accountability, like you and I avoid genital herpes, we wouldn't have gotten past my first reading in 1997. I had just gotten stung by X-2, a consummate, irrationalist, con artist, before I met Carol so my etheric radar was working pretty well by then.

To discover a psychic like the refined ones, including Carol, who look after this network, is like finding the Philosopher's Stone or Aristotle's 'honest man.' I think most of the good ones, and even many potentially powerful but undeveloped psychics, including children, have been murdered in recent years, as I said before, by the various sewer rat agencies. Apparently these agencies also realize that it's getting less and less possible for anyone to sit on the fence these days, so they cover their bases by eliminating even potential adversaries. This is a very, very old occult/corporate practice, though. Clear precedents are the witch hunts that were conducted throughout the previous millennia, by masons within the Protestant clergy, since the early 1600s and the Inquisition conducted by Satanists within the Church of Rome's hierarchy, who burned five million women at the stake in Jesus' Name.

Jeff McKinley, who was subjected to a lot of assaults on his heart by psi adepts, and electronic weaponry and poison in his earlier gifting crusade, told me that he can now easily distinguish a genuinely distressed heart condition from the artificially induced ones. He eventually stopped having genuine heart trouble, thanks to the zapper and orgonite. As soon as he figured out how to counter the artificially induced problems, his heart stopped giving him trouble.

For some reason, the heart and brain parasites die quickly, in the first zapping session, and the heart repairs itself very fast. Intestinal parasites take up to a week to be killed, by constantly wearing a zapper. When 'heart specialist' serial killers, (MD surgeons), open a heart and find fluke worms, they simply sew it back up again and keep their mouths shut. This is well known among the serial killers and their entourage, in operating rooms. Any zapper disintegrates all of those big worms, anywhere in the body, in twenty minutes.

The reputable psychics who are left alive now, are the few whom God has felt pleased to preserve, or who are particularly adept at defending themselves and healing acute illness. Recently, though, some of the psychics who have worked for the bad guys in an altered state, are waking up and contacting us, for help and encouragement. In that case, they've wanted out of the system for quite awhile, but had simply felt powerless, before they heard about what we're all doing successfully.

D Bradley, who had been one of Shirley MacLaine's youthful psychic coaches, in the 1980s, later left the Great White Brotherhood's guru fold, in the mid nineties and brought a lot of vital information, help and encouragement to us all in the nascent stage of this grassroots network's growth and consolidation. His first hand, timely psychic intel helped Carol and I stay on track and out of mortal danger, in the earlier years, and we were able to return the favor before long. He's mainly focused, now, on preventing the bad guys from dragging his young sons into the Monarch Program. Their Merovingian bloodline makes them prize candidates, unfortunately.

The Illuminati bloodlines like Bradley's are considered essential to the survival of the occult/corporate world order, and they don't turn these families loose as easily as they give up trying to harm you and I. Please send him plenty of energy whenever you think of him, okay? You and I will never experience the level of pure horror that he contends with, by going directly against these mega parasites on a personal level and for the sake of his sons. Under the circumstances, he's got more autonomy than he's ever had before, and this will pass as soon as his former masters are fully exposed and dealt with, under the law. He told me that he spends a large part of every day taking down predators now. Carol and I feel that nobody does it better than he does and I'm sure Lawrence Rockefeller would agree, if he were still around.

It's possible to break free and become genuinely empowered, if you're one of those psychics who are still in occult/corporate harness, but getting free of these agencies is a real battle, mostly psychological, especially if you're hooked on pot, gratuitous sex, alcohol or any other spiritually-degrading activity. Our personal progress is always a function of how well we are willing to demonstrate integrity. Nobody is exempt from that responsibility.

The mind seems to be the favored territory of psychic parasites, and this old world order is essentially parasitic, after all; not truly powerful. The mind is also the realm of delusion and the tyranny created by this old hierarchy is mostly imaginary. Their Theosophical, 'irrationalist,' ideology is merely hypnotic, not substantive.

The heart is unknown territory to them, which is why our blasts from the heart are so effective at neutralizing them. If you're in their thrall, you probably got there because they successfully manipulated your ego. The ego boost, in that case, was your programmers' sucker punch, after they systematically and sufficiently insulted, traumatized and dishonored you as a child. They've never been able to force people to work for them, though, and I think this is one of the Big Secrets that is their ultimate undoing. There are countless millions of us who went through this horror as small children, and it's always tough to face that realization.

If you were victimized and can assimilate that fundamental realization, then choose to move into self-empowerment, finally, you'll recognize that the old world order is rather characterized by this slightly paraphrased quote, with apology to Ralph Waldo Emerson:

"When a resolute young fellow steps up to the great bully, the world [order], and takes him boldly by the beard, he is often surprised to find it comes off in his hand, and that it was only tied on to scare away the timid adventurers."

Of course, if I end up getting suicided in coming days, the laugh last might be on me, but I guess I'd be laughing from the next realm, because I've personally caused so much irreparable damage to their enslavement and genocide agenda, that my life would be a relatively cheap price to pay, by now, for all of that job satisfaction.

I want to be safe and prosperous, but I also want the same for you and for our progeny, which is why I took this stand, in the first place. The hackneyed phrase is true, by the way:

> "If one won't stand for something, one will fall for anything".

If you're one of the PJ folk, whose mouth-breather worldview is conditioned by the What To Think Network, this book will merely be entertaining for you, at best. If you, like us, are interested in severing the mind-numbing, Gordian knot that is the potted, popular paradigm, you're going to get some real spiritual weapons and ammunition from this grassroots effort, but that will only begin to happen for you if you're sparked to do the work we're all doing. I wish to spark you with my writing.

The very schizoid 'materialism' paradigm is proof positive, that 'common sense' and 'conventional wisdom', are oxymorons. The diadem of this parasitic world order, (the long-dreaded Novus Ordo Seclorum), is western democratic liberalism, which grew like a fungus out of the faux conflict between Tweedledee and Tweedledum, in the 1940s, and the ensuing cold war, smoke-cured in the sixties by pot-addicted, hammy, faux-rebellious, intellectually larcenous Marxist professors in all of the State/corporate-funded universities and colleges, throughout North America and Europe. That was a fake revolution, that did nothing but further decimate the nuclear family, and expand and consolidate the centralized power of this ancient, patient, patriarchal occult/corporate hierarchy.

Are there any private colleges left, that still teach the classics, which honor the concept of right and wrong, and tell about the essential order of the universe? I think we need a renaissance in education. When that's come to pass, good people like Maxwell, Leibniz, Schiller, Reich, Tesla, Rife, et al, will finally get the credit they're due, and the current parade of academic poseurs, including Marx, Darwin, Newton, Russell, Malthus, Bacon, Einstein, ad nauseum, will finally meet a suitably inglorious, but quiet end.

The What To Think Network and institutionalized programmers, (most clergy and university professors), essentially own the minds of the Pajama People. I include in the PP class, the millions of hang-fire, Volvo-driving neo-activists, who are unblinking devotees of Rockefeller's dire, and colorless Corporation for Public Broadcasting. The mind is a marvelous etheric computer, but it's also the realm of delusion, especially when the fellow in the driver seat is conned into smoking marijuana, or subscribes to the What To Think Network.

We're mainly concerned about the dynamics of the heart, in this movement, as I mentioned. The heart is where discernment, commitment, wisdom, integrity, empowerment and accountability are.

James Hughes, pre-eminent pioneer and exemplar of the

etheric healing arts, reminds his students that the ancient Egyptians couldn't figure out what the brain was for, so when they were putting vital organs into canopic jars, during the embalming process, they just threw the brain away. Maybe they tanned leather with it ;-)

I hope that my own brain is more useful than that, in the long run. What's your brain good for?

I think that a good first step anyone can take to start to break free from the What To Think Network's depressing, hypnotic hammerlock is to read Gary Allen's "NONE DARE CALL IT CONSPIRACY". It's a good, sans-ideology primer, for learning how the Machiavellian old world order and its peniform tentacles have operated, throughout human history.

For the very discerning, maybe absorbing the information in that book will provide enough incentive to start pulling down the remains of the enervating, old-fart patriarchic paradigm. There are many, many other top-notch research journalists, too, who sell their well-documented wares in the open market, without insulting your intelligence by using information slightly out of context, in order to preach an ideology.

Before the internet came along, many of these books couldn't be readily found in the US, due to federal censorship. If you think America's too free to have censorship, please think again.

The US Government actually held a public book burning, as late as the mid 1950s; after all, when they rounded up all the literature they could find that favorably referred to Dr Wilhelm Reich or his findings. After that, I think they just quietly censored certain authors, instead of imprisoning and murdering them, which is what happened to Dr Reich. Fascism is never pretty, after all, even when the trains are all on time.

During World War One, Secret Service agents, including J Edgar Hoover, stormed the home of Michigan's Governor Lindbergh, threw all of his books on the lawn, and lit them up. Hoover punched little Charles Lindbergh in the face, when he tried to interfere with the process. When the Governor had been a US Senator, he vigorously opposed the fake Constitutional Amendment, which handed the US Treasury to a foreign corporation in 1913. Hoover was later rewarded, when the Federal Reserve Corporation created the FBI for him. Federal Police are unconstitutional, of course. Treason was Hoover's primary crime, not the occasional assassination plot; cross-dressing and having sex with men wasn't a real crime, as far as I know.

It's better to buy these books, than to try to get it all for free on websites as someone else's, (including my), digests. Most of these internet sites are disinformation sources, after all. Someone has made "NONE DARE CALL IT CONSPIRACY" available on the web for no charge, and it's a good idea to print it out and read it carefully, if you're new to the arena of critical, creative thinking: http://reactor-core.org/none-dare.html.

Mr Allen is long dead, so you won't be undercutting him by getting this book for free, don't worry. As far as I know, he died as a fairly young man, shortly after publishing a detailed, documented expose on the institution, which the Rockefeller cabal established and financed, to undermine all of the State Legislatures in America, through an extensive, London-financed bribery, extortion and assassination network. That institution will continue to poison our state legislatures, until the Federal Reserve Corporation has been destroyed, which may happen pretty soon. After that, many thousands of filthy state politicians, (nearly all of them are lawyers—go figure), who have enabled the perpetuation of Washington, DC's treasonous government, will be exposed to scrutiny and can be dealt with appropriately, by expeditiously-elected local Sheriffs and judges.

I bet you thought, as I used to think, that these state legislatures were the milieu of useless eaters, at best, and comical petty-criminal fraternities, at worst. They look incompetent because they avoid making any decisions that aren't sanctioned by their occult/corporate benefactors. They're all competent enough to grow fabulously wealthy, on bribes from the Rockefellers' 'lobbyists,' please note. What's cute or funny about that?

Most of the 'active' supporters of these diminutive Foghorn Leghorns are—you guessed it!—the Depression Babies; the first genuine, fascist generation in our nation's history. They were literally born into Roosevelt's Raw Deal, and they remain National-Socialism's staunch defenders, seconded by their cynical, neo-hippie, 'public-sector-employed' offspring; my generation.

Ironically, the Depression Babies believe that World War Two was a just war, rather than a managed conflict, that was designed to spread and consolidate National-Socialism. That happened immediately throughout North and South America, industrialized Asia and Europe, as soon as the war stopped and 'international socialism' was instantly set up in every other country that had sufficient infrastructure to support it.

These oldsters are quickly dying off now, as I mentioned before, because subscribing to drug-based, socialized medicine, (another Big Lie), is suicide by installment, overseen by many thousands of Kavorkian wannabees. Better hygiene and infrastructure tends to extend the human lifespan, but the global medical/drug cartel tends to ensure that anyone who lives past seventy, will gradually die of cancer and other costly, induced and managed illnesses, leaving their progeny with very little of their parents' acquired fortunes. Looking back, it's not hard to see how the National Socialist regimes induced relative prosperity, among the Depression Babies, with the full intention of re-absorbing it all through their extensive Hospital Gulag Archipelago. Taking the long view has been essential to expanding and consolidating tyranny, but they apparently never saw zappers and orgonite on the horizon, until it was too late to stop this liberating technology. Or else, they were too arrogant to admit to themselves that something this simple could be a threat to their hegemony.

Centralization is the current political paradigm, and if there's going to be genuine government in the world, as this one degrades and fails, it will be up to the younger generations, (the ones following mine—most of my generation are too inebriated and complacent to do much of anything), to make it happen from the local end, first. We just don't need any more intrusive, overbearing centralized government—we never did need that, in fact.

When the treasonous, over-centralized national governments of the world fall, under their own excessive weight, (and Chinese fiscal intervention?), in the perhaps not-distant future, their hidden corporate 'influence' networks will suddenly evaporate, and we'll have genuinely representative state, local and county governments, for the first time in living memory and probably by default, so why not get ready to make your vote count, in your own community, when that comes to pass? All of the communities on the planet are intimately connected via the internet, so this will be a global groundswell; unopposable and grassroots by nature, just like this expanding gifting network is.

If you'll toss out a simple, three-ounce, orgonite Towerbuster in every block in your town, it will probably ensure that the electorate will be attentive enough to choose reputable people, for your local government, after the fall of the Fed.

If reputable people are chosen at the local level, they can network well enough with other local governments, to effectively pick leaders for regional governments, which will have as little power as possible and whose representatives will draw suitably modest salaries, not suck on the corporate sow's teat. This can be extrapolated all the way to the global level safely, and it's going to happen in an organic, timely way, as the internet happened in the previous decade.

For a couple of generations, after the American Revolution, the US Congress had a lot of good, genuine people in it, like Daniel Webster, who did their best to limit the expansion of federal power over the states. Maybe that was a little ahead of its time but now is the time for their vision to come to fruition. The only relevant characteristic in elected leadership should be genuine self-sacrifice, and the role of President should be no more significant than the role of a small town mayor.

It's always been curious to me, that some people consider 'globalism' to be inherently evil. It doesn't need to equate with centralized power. The internet is showing us that a properly low profile, and non-intrusive global infrastructure can help everyone, and the small size of its administrative staff guarantees that it can't be intrusive. The names of the people who own and operate the internet are available, but most people simply don't care who they are, because the internet serves everyone well and is reliable. We can derive a lot of hope for real government from that simple, working paradigm.

The internet does extremely well without any corporate-teat or government interference, and it's a viable, vital global network. So we can assume that genuine world peace and prosperity are attainable right now.

In order for that to happen, of course, local and state/district/

province governments have to become more autonomous, right away. Excessively centralized government won't willingly give up any of its power, which is why all of the top-government elected and bureaucratic officials need to be held accountable. Most have obviously committed treason in that. The only relevant way to approach the average US Congressman or judge these days, rather than with a petition, is with handcuffs.

What could be more natural than the unfoldment of this process, once enough people have consulted openly about it? Are you afraid to utter the word, 'treason?' to people you know? Let duly lawful courts, which can be quickly set up under the rule of law, deal with the mavens of the parasitic, corporate world order at the proper time. There is abundant evidence of their crimes in the public record, after all.

JFK tried to achieve this on a national level, from the Oval Office, and maybe it just wasn't time for that, or maybe it's not a function of government to hold itself accountable, since excessively centralized government, including the unlawful JFK White House regime, perpetuates itself.

What will it take to help you break down your programmed feelings of hopelessness and powerlessness, if that's what you're feeling at the moment? When you've gotten involved with this unorganized network, you'll learn some tricks that many innovative people have discovered or developed, over the past five years, to induce critical, creative thought.

Our public enemies apparently assume that discrediting us with slander, or shifting attention away from us and toward the fakers, can slow down the expansion of the network but, as I mentioned, the other side can't understand heart-oriented movements, so every attempt they make to sabotage the network ultimately serves us, instead.

Aliens, these days, are practically a non-issue, though if/when we start seeing them in the streets, it will change the way they're spoken of, or ignored by our critics, I guess. I don't consider alien, parasitic intervention to be more than symptomatic or opportunistic. That's why I rarely mention the subject. We can assume that disabling parasitic human institutions in appropriate, and constructive ways will solve the worst of humanity's problems, in the short term, and the opportunists have been leaving our world in droves already, so why angst about them?

Having personally seen aliens and their ships, I don't consider them any more or less supernatural than Chinese or Kalahari Bushmen are. I encountered some Bushmen in Namibia, and they felt a lot more exotic to me than reptilians do. The Bushmen's presence felt more like I was with dolphins or Sasquatch, than with people, in fact.

I think Marco Polo and the Chinese who viewed him, underwent a bigger culture shock, than any of us have, by encountering aliens or reptilians lately, in fact. All that Marco Polo brought back with him, as far as we know, were noodles and sauerkraut. If I'm not mistaken, orgonite and continued help, protection and guidance for us all—and by extension, for humanity—have come from The Operators, some of whom are no doubt aliens.

Dr Reich, who had established scientific standards, which we've been attempting to live up to, originally built his cloudbusters in the late 1940s, to fend off energy-poison attacks against his lab facility in Maine, from ships outside of our atmosphere. Those DOR beams were killing everyone and everything, because the massive orgone accumulators on the property were collecting and condensing that energy, and then radiating it back out. Using the cloudbusters for weather modification, after he destroyed and/or drove away the spaceships, was serendipitous, as were many of his other major discoveries.

I'm sure that if he had known about orgonite, he'd have adopted it in favor of orgone accumulators for many applications, but catalyzed resin, which is what most of us use as the organic matrix for our orgonite, wasn't popularly available until around the time of Reich's imprisonment and subsequent murder, in the mid 1950s.

Looking back on my journal reports in the first few years, Carol and I can see some truths that weren't apparent to us at the time, but those written accounts in my internet journal, are all still fresh and lack hypocrisy. Since I'm happy to admit that all of our accounts are merely subjective, our public enemies can't get us squarely in their gun-sights, so when they slander us or try to discredit us, they're only giving us free advertising, instead. If I had agreed to be a 'leader of the movement' or had otherwise been successfully ego-baited, they'd be having their way with us by now, and you wouldn't likely have ever heard of orgonite.

For example, a couple of years ago a charismatic self-seeker, who wears Christian fundamentalism on her sleeve, and claims to be God's Mouthpiece told me that, 'Bible code clearly shows that Don Croft is a prophet and world savior!' I felt repulsed by that comment, and politely declined her offer to promote me 'among all the Christians.' My sense of revulsion was confirmed, a few months later, when that 'Bible code' software program she was pushing, came under closer scrutiny and was found to be fake. Roger Bacon apparently turned the Bible, (the King James version), into a Masonic codebook, during the time he was overseeing the slaughter of women in Great Britain, as a top mason.

That gal instantly became one of my most vehement public

critics, (she also started hammering Carol for being a witch), and now she's got a fancy disinformation site, that is apparently designed to draw attention away from our energy-tossing successes. See how the disinformants' job descriptions get altered, now and then, to fit the other side's mutating sabotage agendas? After similarly being discredited, the 'sylph spokesman' became a featured 'orgonite expert,' for instance. If you'll pay attention, you'll see this tactic used, again and again, by the other side.

Dr James DeMeo, a close companion of Dr Eva Reich, and our most enduring and outspoken opponent, traveled to Germany in January 2004, to warn his audiences not to experiment with orgonite, especially 'Croft-style cloudbusters.' Of course, as soon as he left, many Germans made orgonite cloudbusters, because of simple curiosity, and they started reporting their consistently dramatic results on internet boards, then many more did the same.

See how it works? The other side seems to be thrown by the divine Potter to continuously sabotage themselves, by attacking our work. Georg Ritschl of www.orgonise-africa.net hosts the German-language internet orgonite forum. Georg offered to publish a companion volume to this one that's more technical—a manual for making and distributing orgonite devices. He hasn't suggested a title, yet, so if you want it, do a search on www.mazon.com using his name, and when it's published you'll find it that way. Meanwhile, www.ethericwarriors.com has the technical information you'll want, but in a less concise form. He's set up to get orgonite to African gifters throughout the continent, through a very generous sponsorship arrangement. In that case, he's literally giving the orgonite away, because your donation only covers his manufacturing and shipping cost. Nearly all Black Africans are kept in penury these days, but many of them are intrepid and avid gifters, who get spectacular results, so your material help through Georg's site is very productive, as well as appreciated.

Some of the German networkers have gone pretty far afield in their efforts, by the way. Richard Eichenlaub and Gerhard Huber went to Spain last summer, with a carload of orgonite and a dozen or so cloudbusters, after Todd Platzer, an American gifter living in Seville, invited them to base their operations in his home. Todd's the fellow who immediately gifted the underground, psychiatric hospital Gulag, in Cleveland, in which Roger Elvick, a political prisoner, was being tortured last year. We went to bat for Roger, as soon as we found out about that, (it had just started, apparently), and Todd gifted the facility the next day. Roger's situation immediately improved after that, thank God. Todd took his family to live in Seville shortly after that.

There's a well-established, international French gifting network, too, also with a lively forum, originated by Steeve Debellefuille of www.Quebecorgone.com. Steeve is a good friend, who also set up the only, (at the moment), viable English-language forum, Etheric Warriors. Jacques Lasselle is administering it these days, and he's a miracle man, when it comes to countering the savage effects of Langley's and Ft Meade's unwashed, hacker hordes.

In Finland this is spreading fast, among youth, due to the tireless, international networking and gifting efforts of Pekka Martinen and his partners, Jukka and Tapani.

Igor Cinti hosts www.orgoneitalia.com in Rome, and has stuck his neck way, way out by thoroughly gifting the Vatican, including disabling a huge HAARP array there. He's suffered some pretty severe reprisals on that count, but fortunately he recovered from each one, stronger than ever. His offerings are well known on the internet, in that country, and the visual and etheric effects of the Italians' collective gifting efforts have been indescribably beautiful. Italians seem to have an exceptionally well-developed aesthetic sense encoded in their DNA, after all.

There are less publicized but very active grassroots gifting networks in several countries, including Argentina, Brazil, Malaysia, Australia, Japan, UK, New Zealand, Uganda, Poland, Romania, Czech Republic, Greece, Kenya, Congo and Sudan. Most people who gift don't participate on internet forums, nor are they interested in organizing the effort.

Most of our African compatriots choose to focus more on orgonite distribution than on internet reporting, though Judy Lubulwa, in Nairobi, and young Abdullah Jim, in Uganda, take a lot of pleasure in sharing their eloquence and insights with us, on www.ethericwarriors.com. Dr Kayiwa sends me concise reports and comments following his international gifting and lecturing trips. Judy is making inroads, now, reversing the previously, horrid ambience of Nairobi, and growing an informal network of enthusiastic gifters there. The psychics see signs of quite a bright personal future and unfolding destiny for her.

My good friend, Dr Rushidie Kayiwa, in Uganda, loads his car with towerbusters and charges into the bloodstained, corporate-terrorized areas in Central and East Africa to create and enforce peace, by simply distributing orgonite there. He's ended near-famine level droughts in Sudan this way, and is starting to manufacture affordable zappers in Kampala for distribution throughout East Africa and beyond in coming months. He epitomizes, for me, the synergistic combination of the combat soldier's aggression with the healer's compassion, which is an example that every gifter ought to live up to.

Georg, who lives in South Africa, has almost single-handedly reversed most of the previously near-famine level droughts in Southern Africa. He initiated sustained and unprecedented rainfall, throughout the vast Kalahari and even part of the Namib, which is the driest region on the planet. He's worn out two vehicles over countless thousands of miles of unpaved, African highways and tracks, on his HAARP-busting and cloudbuster distribution missions, so far. Www.orgonise-africa.net is his site, as I mentioned above.

Though it may be too early to tell, Carol and I feel that the Japanese may suddenly outstrip us all, thanks to the timely pioneering efforts of Tetsuzi Moriwake, my Japanese brother in Hiroshima. I think it's poetic justice that the publicized part of the gifting movement in Japan should start in that city. Ed Schindler in Kyoto, ('Eddie-san'), is the man who inspired me to 'invest in people', by his selfless example, and we were privileged to meet him, when he came to the US last year. He and Kelly McKennon visited Tetsuzi-san last summer and they created some spectacular effects in the atmosphere there, inducing an immediate visit from some sylphs, which Tetsuzi-san photographed and posted on Etheric Warriors.

The Canadians, per capita, are the most numerous nationality represented in this grassroots network, and it was Steeve DeBellefuille in Quebec, who facilitated the establishment and growth of the gifting network in France, and into Switzerland and the French speakers in the Benelux countries. Steve Baron in Toronto, initiated and coordinated the network's most extensive and comprehensive urban gifting campaign, second only to D Bradley's massive, successful atmosphere/ambience transformation effort in Los Angeles.

I'll be surprised if DB's personal effort will eventually be matched by another individual, but Steve's claim to fame was rightly earned by his phenomenal networking achievement.

Many people, in and around Toronto were involved in making and distributing tens of thousands of orgonite and scores of cloudbusters throughout that huge metro area and beyond, within two years, and the number of genuine activists there is growing.

Another Canadian, Dave Emmett, has committed to gifting the entire Caribbean from Barbados and Jeff McKinley has been instrumental in supplying cloudbusters for that effort. Participants have been volunteering on some of those islands by now, and the number will grow.

I get a kick out of the way AOL preaches 'safety from hackers', because AOL is owned by the CIA, and is a primary surveillance tool in America now, for the Homeland Security Abomination. When you log on to AOL, you're treated to a bombardment of subliminal programming and frequency blasts, that might even exceed television's and the quasi-official disinformation sites, that abound on the web. The 'free' AOL subscription CDs that are available in stacks, in US Post Offices, give off so much DOR that psychics prefer not to stand close to them. Concentrating deadly energy on CDs must have been quite a technical coup, for the sewer rats.

I suspect that nearly all of the hackers who routinely crash the computers of people who visit sites like ours are on NSA, MI5/6, KGB, Mossad, NSA and CIA payrolls. AOL enables that, rather than shields against it, by 'caveat emptor.' I once visited a disinfo site, when I did a search for 'time travel', on someone's AOL-hosted computer and the hard drive was destroyed instantly. Their latest answer to 'security' is apparently to prevent their subscribers from visiting genuinely informative websites, or from getting email from people like me.

There are essentially no predatory hackers out there,

except for the ones whom the NSA and CIA employ or have dissociatively programmed, after all. To believe otherwise is a mind control coup. The loose-cannon Monarch asset hackers are simply turned loose on the populace, otherwise. Most of these folks don't have the sense to create a life for themselves, but they're easily aimed at you and I, by their handlers.

A serviceable firewall notifies you, when ordinary hackers and most professional hackers try to get in. Because I only do my mail on the web, I haven't used an anti-virus program for several years. I haven't had a problem with crashed hard drives, either, since I stopped using email software in January 2003, and you can guess how much the Feds would like to shut me up online, even temporarily. Carol and I both have online laptops, and we keep a spare, just in case.

I hope you're getting a sense that everyone in this network is doing this work because it's empowering, and eminently satisfying. This is a genuine grassroots network, not a personality cult or dues-demanding, lock-step organization. I'm certainly not in charge of it, though I do exploit a bully pulpit, when there's trouble to deal with. That doesn't happen much.

If you're a person who has a social conscience, you'll no doubt try gifting, since it requires very little time and material investment, but is said by so many, to reap very large rewards. If you're one of those rarities who have a social conscience and can accurately feel and/or see subtle energy, you'll get more immediate confirmations. Otherwise you'll need to do some legwork and intelligent orgonite distribution, before you'll get dramatic, visual results in the lower atmosphere and sky and a sudden drop in predatory behavior and angst among the population.

If you hate tyranny and have the killer instinct, which shouldn't be seen as a bad quality, you can do your share to neutralize predatory hierarchy by creating or joining existing cells, to toss energy through your heart at specific, identified mass murderers. These 'predator safaris' are quite effective and gratifying.

We develop our confidence in orgonite's phenomenal etheric potency, through direct experience. If you'll do enough gifting, starting with your home, neighborhood and workplace or school, you'll witness immediately-improved attitudes and behavior among your neighbors, family, fellow students and coworkers. After that, we can openly talk about the magic-bullet aspect of simple orgonite, in terms of practical, grassroots political and marketplace reforms.

> A renowned teacher was once asked, 'Why do you repeat yourself so much?' He answered, 'What did I just repeat?' and the student said, 'I don't remember.'

'That's why I repeat myself,' the teacher said.

I mainly want you to do your part to change this planet into a paradise with your own simple, inexpensive orgonite devices, and conscientious, appropriate energy tossing. This notion might still seem foreign to you, because it's so new, so I've repeated the request in many interesting ways, (I hope), so that you'll respond by achieving your own unique exploits.

It's the personal process which counts, not the comparative achievements. As it's been said, everyone has different capacities, interests and ambitions and all of that is appropriate. Whatever you believe you're capable of, you're about to find that you can do, see and learn an awful lot more than that, if you'll seek to demonstrate personal integrity and exercise your social conscience, by applying the information that's available from this network.

Chapter Eight

Be Well!

Healing the earth for the duration is feasible for the people in this network, because we're able to keep our bodies free of serious illness. That's mostly made possible with zappers, which destroy all of the living pathogens in the body and neutralize most of the chemical pathogens we encounter—even rattlesnake venom, (I know that from personal experience). I'm sure that if it weren't for zappers, many of us in the US and elsewhere would have been murdered by secret police agencies with poison, electronic weaponry or biological/chemical/radiological weapons. This notion seems bizarre to anyone who subscribes to the What To Think Network, of course, but the consistent record of posted experiences, by reputable people, on the internet over the past five years, bears witness to this simple fact.

Biological and chemical weaponry, which usually induces fast cancer, in targeted enemies of the present day order, are easily and fairly quickly neutralized by any zapper, but when one of us gets a potentially fatal dose of toxic metals, we've been getting Dr von Peters' thorough and able chelation help. Not even the Terminator zappers will remove that, though it's pretty good at getting smaller doses out of the body. Fortunately, only a few of us in the front rank have been heavily dosed with toxic metals, including beryllium, lithium, lead, mercury and tin. The problem for the sewer rat agencies is that they apparently have to administer doses that are small enough to kill us gradually; otherwise it would look too obviously like murder.

Orgonite close to the body seems to neutralize radioactive materials in the body. I first noticed that eight years ago, when

I got some feedback from a zapper distributor in Romania, who inadvertently healed a fellow that was dying of lung cancer, induced by exposure to radioactive material, while cleaning up Chernobyl. He still had quite a bit of radioactive material in his body, then, and all trace of that was gone, when his near-fatal lung cancer was completely cured, after a few weeks. The zapper that he used was one of my first ones that included orgonite, which I'd just learned about and begun experimenting with.

Cancer cure testimonials are quite common with all zappers. I think it's the most reliable, fastest, easiest and cheapest way to cure cancer. The curative component of any zapper is a circuit, which you can build for a dollar's worth of wholesale parts, so this shouldn't be mystified by anyone. Earlier zapper makers traded on people's credulity by selling overbuilt, expensive and partially effective zappers, and promoting a boatload of fake science and groundless recommendations and warnings but, thankfully, most prospective zapper customers, now find their way to more legitimate sources.

Lots of people are making and selling good zappers. Many of these zapper manufacturers' sites, including ours, are listed on the homepage of www.ethericwarriors.com.

The comprehensive Tissue Mineral Analysis test, which Dr von Peters administers through the mail to determine the body's needs, is phenomenally accurate; in fact the brilliant pioneer who developed this test was one of hundreds of casualties, of the generations-old, federal 'War on Healers.' That's an earned distinction that nobody wants, of course, but when the federal government destroys the life of any healer, it's a posthumous endorsement. Very few serial killers, (MDs), are ever prosecuted, though they kill millions of people annually, as the public record shows. How will future historians explain the peculiar way that MDs, politicians, bureaucrats, et al, get rewarded for their grand-scale failures and crimes, while genuine healers and researchers routinely get killed and imprisoned? If you've read this far and still wonder why your doctor or the What To Think Network never told you about zappers, then you obviously haven't been paying attention. What I'm telling you in this chapter can easily prevent you from ever getting seriously ill, or from dying under the knife of a sanctioned serial killer. OO7 wasn't the only fellow licensed to kill.

In addition to his unique, effective and gentle herbal chelation technique, Dr von Peters' treatments include specific, 'Homeo-Herbal' remedies of his own design; also appropriate nutrients and minerals for re-establishing balance within the body. For many years, he's specialized in curing 'hard cases:' the chronically ill, whom MDs and other naturopaths have given up on.

At our request, two years ago, Doc von P formulated ChemBuster for use with zappers, to help people repair their damaged vital organs, in cases of the new chronic illness endemics that were mostly induced by chemtrails, before the middle of 2002. You can read about ChemBuster on www.lifequestformulas.com or on our site, www.worldwithoutparasites.com. It's an inexpensive, Homeo-Herbal remedy that usually shows positive results in a few days, after one has reached a healing plateau with just zapping. Zappers are mainly curative devices and ChemBuster and similar remedies are healing products. Curing and healing often need to happen together. The Doc was astonished by the overwhelmingly positive results of this remedy, because he's always felt that the 'magic bullet' approach is misleading, at best. He discovered a manufacturing method that causes his remedies to pick up ten times their natural orgone energy, when placed next to orgonite.

Carol and I, also several other pioneers within this movement, whose bodies were put out of balance with poison and energy weaponry, have gotten critical, timely help from Dr von Peters. There are probably many viable ways to get the body back to vibrant balance and health, but Dr von P guarantees good results, even through the mail. How many doctors do you know or have heard about, who can make a guarantee like that?

His sites are www.uncurable.com and www.lifequestformulas.com. Dr von Peters re-established the First National University of Naturopathy system, of which U.S. School of Naturopathy is the second, nationally chartered, naturopathic school in the U.S., established in 1910. FNUN is the first university system of natural medicine, and the Doc is the most extensively credentialed naturopath in the U.S.

He's titled 'Professor' in Russia's academic system, which is second only to 'Academician.' He lectures in Russian medical schools regularly, and brings back the world's finest electro-medicine healing devices, including the Pro Vitalizer Plus, which he sells on www.lifequestformulas.com. These are fairly expensive, but discounts are available in quantity if you contact him from the site. A detailed, easy to follow therapeutic manual is now available in English. Carol and I have and use one of these miraculous devices. They're capable of inducing profound healing effects in the body, when used properly, and this device has helped us quickly recover from frequent electronic and ritual magic assaults, on our hearts and energy fields.

It costs about the same as a Rife generator, which is only effective for skilled dowsers. The reason most Rifers don't get good results is that the frequency tables have been obsolete for decades, so only skilled dowsers are capable of finding appropriate, resonant frequencies now. It's a peculiar feature of human nature, that a lot more people will gladly spend that amount on something that hasn't been shown to work properly, but is sufficiently hyped, than will spend it on a device, such as the Pro Vitalizer Plus, (previously known as the RussTech DT), which has been abundantly proven to get phenomenal, consistent results, but gets very little play in public forums. Dr Clark's plagiarized version of Dr Rife's frequency medicine protocols is simply a hi-tech zapper, of course. Dr Rife got his astonishing results by applying frequencies through a flashing light, not through electrodes applied to the body. In the case of Dr Clark's protocols, more results can be attained with a cheap zapper circuit than with a bulky frequency generator, because at least the zapper can be worn, in that case.

It's tragic that Dr Rife was a world-class pioneer and scientific genius, but that present-day poseurs and incompetents promote his useful invention, out of context, and produce no discernible results. The same sort of chicanery was happening in the zapper industry, until we turned the tide in recent years. I hope someone will eventually do for that great man what the orgonite movement has been doing for Dr Reich's savaged reputation.

I hope that this book will generate some well-deserved attention for the new, fascinating Russian electro-medicine devices, because it's been a hard sell for Dr von Peters, in the face of overwhelming fake science from the likes of the ersatz Rifers, who waste valuable space in alternative healing forums, these days, and drown the mellow voice of rationality, with a cacophony of groundless sensationalism.

The few people in this network, who mainly get assaulted and poisoned by quasi-governmental sewer rat agencies, are the ones who regularly participate in the international, group 'predator safaris', in the Etheric Warriors chatroom, or are doing exceptionally good and extensive gifting work. You're not likely to get hammered just because you disabled a few hundred death transmitters and HAARP facilities in your area, or etherically smacked down a few Homeland

Security Abominations who killed your dog, or burglarized your home or office, don't worry, probably because there are so many activists at that level that the Homeland Security Abominations and other two-legged sewer rats simply lack the personnel to harm or thoroughly surveille us all, so they focus their resources on those of us in the 'front rank', on the etheric battlefield. It seems to take them a lot of man-hours to line up even a minor assault on an individual. They probably wish they could just round us all up or shoot us in our beds, and have the What To Think Network call us 'drug dealers.'

Two of the few people who have remained faithful friends and allies with us, through some challenging years are Laura 'Dooney' Weise and Dr Steve Smith in Western Montana. Dooney has been training people to toss energy effectively for over a year and Dr Steve, her husband, operates a successful healing clinic. Both are competent, reputable psychics and enjoy remote healing and predator-hunting with groups.

Dooney is able to remotely monitor a student's blasting technique and give pointers in real time, until the student succeeds, after which he/she usually feels confident to take on predators alone. Every bona fide healer in this world is guaranteed to face occult/corporate predators' reprisals, of course. If you think you're a healer, but are not under assault, it probably means you're off-target, in fact, not that you're 'spiritual.' Maybe before long, the serried legions of psychic predators, (mostly 'off-duty'—read: altered state; out of body—Europoid and Asian new-agers), who lie in wait to aggressively interfere with healers, will be held accountable under the law, before long. After that, we won't have to use their reprisals to measure our effectiveness. Dooney charges a nominal fee so that she won't burn out, because teaching this new discipline is often hard work. If you want these lessons, contact her at dooney@bluemarbleimages.com and she'll let you know how to get started.

After Carol got poisoned in Hawaii last month, Dr Stevo sent her a new 'peptide clathrating agent' immediately and she recovered in a day or so, as I mentioned earlier.

> Peptide: two or more amino acid molecules linked in a chain.
> Clathrate: a compound in which molecules of one component are physically trapped within the crystal [chain] structure of another.

The clathrating compound in this remedy, effectively seeks out toxic material, including heavy metals, toxic chemicals and mycoplasmas, literally traps them and moves them harmlessly out through the urine, feces, skin and breath. It's easy to take: just a few squirts under the tongue, twice a day. I was pretty astonished to see how rapidly Carol improved, and we'll keep this in our first aid kit from now on, along with the clay tablets, that Steve provides for moving toxic material out of the intestinal tract, quickly and safely. As with any competent physician a full consultation gets the best results, but there are a few good remedies like this one, which can get us through crises. You can order this product from Dr Steve Smith at drsteve@4quantumhealing.com.

Ordinary chelation techniques leave a trail of destruction, in the process of stripping toxic material from the body, but the above-mentioned methods are entirely harmless and beneficial. This isn't glitzy techno-babble; it's an organic, dynamic process that works.

It may be that simply applying vitalizing micro-current to the blood regularly, has kept all of us from falling prey to the various assassin agencies. Nicola Tesla might concur, because in 1888 he wrote,

> 'I was so blue and discouraged in those days, that I don't believe I could have borne up, but for the regular electric treatments which I administered to myself. You see, electricity puts into the body what it most

> needs: life force; nerve force. It's a great doctor, I can tell you—perhaps the greatest of doctors.'

Appropriately designed orgonite pendants, worn near the heart, can block directed energy weapon attacks, apparently, from satellite platforms or from other craft above the atmosphere. We know that Carol's Harmonic Protectors work, but the other vendors who make orgonite pendants are responsible for gathering their own corroborative testimonials and observations, in order to make that claim. We're loath to make exclusive claims, but we have to earn a livelihood, too, so we promote our products in a fair, rational way, as anyone should.

In both of the extreme cases of energy weapon attacks, (that we're directly aware of, at least), the green grass within a radius of ten feet of where the intended victims stood, turned yellow within an hour or so, and the instant assault was accompanied by a blinding flash. That happened to D Bradley in Los Angeles in July 2004, then two months later to Carol in Moscow, Idaho.

We don't know how it was done. I think a scalar or microwave assault would have lasted much longer. Carol was knocked down at least twice, by apparently scalar assaults on her heart from fairly close range, but when the two instant assaults happened to Carol and DB, the energy came from overhead and there were no planes or helicopters in sight, so we assume it was done from at least one satellite. An owner of one of the first cloudbusters reported a similar incident, when an energy beam was apparently aimed at her cloudbuster in July 2001, but that time the bright flash was accompanied by a loud explosion sound.

We visited Bradley a week after it happened to him, and he'd gotten quite ill after the attack, but not debilitated and he was slowly recovering. He showed us the patch of dead, yellow grass on his irrigated lawn, where he'd been standing in that moment. He wasn't wearing any orgonite at the time he was assaulted, but there was a lot of it right around him in the yard and on the fences—his orgonite factory was his backyard in those days—so maybe that absorbed some of the force of the attack.

Carol was walking in the cemetery in Moscow, Idaho, the afternoon it happened to her, a month or so later, and she felt almost nothing, nor did she develop any symptoms, though she felt a little pressure. The flash apparently distracted the driver of a car that was passing nearby on the highway, because he immediately drove, at full speed, right into the deep ditch, beside the road and there was no oncoming traffic. She was wearing a Harmonic Protector. She took me to that spot the following morning and I saw the patch of yellow grass where she had been standing, surrounded by well-watered, green summer lawn.

According to sensational 'Chicken Little' disinformants, like Col. Bearden, Jeff Rense and Sorcha Faal, the entire human race is under the threat of imminent annihilation, by a combination of acts of God and a terrifying nemesis made up of powerful, irresistible scalar and other arcane weaponry. If anyone were to be targets of anything less than an act of God, you can assume that the people in this network, by now, would have been primary candidates, considering how much damage we've all been inflicting; many times more damage to the bad guys' predatory infrastructure around the world, than any other group or network has done. None of us have been critically harmed, in the five years of the network's existence. Here's a fun, edifying discernment exercise, if you're still susceptible to disinformants' enervating prattle: compare the following, which I copied from the script of the movie, "MYSTERY MEN", with any of Bearden's indecipherable but fascinating articles or talks.

> "The psychofrakulator creates a cloud of radically fluctuating free-deviant chaotrons, which penetrate the synaptic relays. It's concatenated with a synchronous transport switch that creates a virtual tributary. It's focused onto a biobolic reflector and what happens is that hallucinations become reality, and the brain is literally fried from within.
>
> "The equations were so complex, that most of the scientists who worked on it wound up in the insane asylum."
>
> Dr Heller

Fascinating and terrifying! Give up hope! Surrender or die!

As I mentioned, inexpensive zappers will inevitably do to the global, pharmaceutical cartel and the hospital Gulag Archipelago what free energy devices are going to do to the petroleum/nuke cartel, before long. A beautiful aspect of this notion is that zapper manufacturing and distribution and, (potentially, at least), free energy device production are cottage industries. Carol and I sell more zappers than anyone, but making and distributing them isn't even a full time job, usually. We work out of two rooms in our home; I do most of the work on the zappers these days, and handle most of the correspondence and Carol does nearly all of the work on her orgonite inventions, which also sell briskly by referral. I started making zappers when I was barely hanging onto my sanity, and living in my car in April 1996. Young Eric Nagal, a gifter in Batac, Phillipines, is manufacturing his own progressive design at home, and distributing them in Asia and the other vendors who are listed on Etheric Warriors all work out of their homes.

Someone is selling, on E-Bay, a ten-dollar zapper, made of an exposed 555 circuit, three resistors, a nine-volt battery and two leads with alligator clips. I guarantee that this one destroys pathogens just as fast and thoroughly as our $134 Terminator zapper can, (our price is subject to increase). The cheapest one, if worn enough, probably does the job a lot better than the outrageously expensive and questionable 'multi-frequency' zappers with timer circuits, or the ones that plug you into house current, because it's easily portable and can be worn under the clothing, like our zappers.

I mostly zap while sleeping at night with a Terminator or ordinary zapper, applied to the sole of a foot with an elastic, terrycloth tennis wrist sweatband. The body repairs itself mostly while we sleep, so why not give it a boost while destroying any pathogens that may have accrued during the day? The body seems to especially love absorbing the vitalizing boost from the subtle energy components, through the sole of the foot.

The reason wearing a zapper gets the job done better than fiddling with electrodes or sitting by a wall socket can, is because the more one wears a zapper while sick, the faster the cure will be achieved, especially in cases of intestinal worms and systemic fungus colonies. Rapid curing is a function of how much micro-current can enter the body, over a given period but we favor very mild current, about five milliamps from a zapper's electrodes, because this won't cause the acids in the blood to rush to the negative electrode and burn a cauterized path up through the skin. This can't happen on the sole or palm at this level of amperage. Those acids are mainly produced by bacteria, viruses, fungi and worms, so when these organisms have been killed, the blood returns to a properly alkaline pH and remains there.

Smoking tobacco, drinking a lot of booze or taking pharmaceuticals makes the blood constantly acidic. I used to visit a steam bath that had cedar benches, fastened by costly monel screws. Monel is the most corrosion-resistant metal alloy for making fasteners and it's usually used for marine applications. No electrolysis was happening, because that was the only metal exposed to the elements, (steam and sweat). The owner of the steam bath was quite incensed by the fact that the sweat from smokers was eating thru those monel fastenings, as fast as he could replace them. Nicotine is extremely corrosive. If it can do that to this metal, imagine what it's doing inside the human body. I bet you thought that tobacco addiction mainly damages the lungs. Organic tobacco, including American Spirit brand, has a lot less nicotine. Standard brand cigarettes have added nicotine, apparently. Another aspect of schizophrenia in our culture is that the US Government subsidizes tobacco companies, while vigorously opposing the use of tobacco. Smokers have become scapegoats similar to Middle Ages European Jewry.

The curative element in the zapper is based on an inexpensive, simple 'on/off' 555 integrated circuit-based assembly, which anyone can put together. Some of us add subtle energy components to that circuit, in order to boost the body's own healing potential. In spite of the boat load of fake science and bizarre caveats and recommendations that immediately attached, like barnacles to the popular conception of zapping, in the mid 1990s, it's really no more complicated than the principle of steady ionization, via micro-current, which is pulsed gently thru the skin by capacitance. It only takes a little electricity to destroy all parasitic organisms in the body and neutralize acidic toxins, including prescribed pharmaceuticals, all of which are acidic poisons. Since these two categories—parasites and poisons—cause most of the illnesses in the world, it means that zappers quickly cure most of the illnesses in the world.

There's nothing complicated or hard to understand about

how this works, in spite of worn-out, strident claims by a few lettered people, to the contrary. A peculiar aspect of modern scientific education is that the graduates are almost completely unable to understand simple concepts like this one. I think they're trained to assume that every process in the physical universe is complex, ultimately indecipherable, and needs to be compartmentalized as much as possible.

If curing illnesses could only be accomplished by something fairly complicated, like a clathrating agent, how could an average, uneducated Joe like me be responsible for successfully curing thousands of cases of acute and chronic illnesses, over the past ten years with cheap, simple circuitry and nine-volt batteries?

I was doubtful about this basic zapper claim, even after I cured my lifelong depression in a few minutes, ten years ago. So, within a month I made and distributed seventy simple zappers to people, who were acutely and chronically ill and they all soon got well, too. I then decided to make a career of this, because it was my first glimmer that a single person can significantly damage a tyrannical global cartel; in this case the ruinous pharmaceutical/medical cartel.

'Curing' and 'healing' are fairly separate processes, of course, just like microbiology and astrophysics are fairly distinct sciences. Sure, there's some blending where they meet but, practically speaking, destroying the pathogens, which cause illnesses effects cures. One still has to heal the damaged tissue and help the vital organs regain a proper level of function, though. According to Dr Rudi Verspoor, curing illness is an act of aggression; healing is an act of love.

Carol and I developed the Terminator in July 2000, right after we got together. We wanted to boost the body's healing capability this way, after it became obvious that a lot of people had become chronically ill from the physical damage caused by chemtrails. These new chronic cases needed much more than just curative help, in order to be well again. I felt sure that there was some way to help these folks heal themselves, other than to just refer them to competent doctors, of which there are just too few to handle the load right now.

Before 1999, I didn't encounter the 'new' chronic illnesses very much, among my potential zapper customers, but I was dismayed that these few cases weren't usually resolved with basic zappers. In the days before chemtrails one mostly heard of three 'untreatable' chronic sicknesses:

'Fibromyalgia' is a catchall term for constant body pain, 'chronic fatigue syndrome,' which is a catchall term that indicates that the body is run down and can't restore itself any more and the sufferer has very little physical energy.

'Chronic Fatigue Syndrome,' or 'yuppie disease,' is just profound tiredness and it's merely the result of the diminished function of vital organs.

'Lyme disease,' started showing up in population centers throughout the U.S. and Canada with a flourish, in the mid-1980s, after a bio-engineered pathogen 'escaped into the environment' from the US Army's biological weapons production facility in Lyme, Connecticut, (which is near Montauk). That episode was actually mentioned a few times by the What To Think Network, and it was right before the term, 'Lyme disease,' was made up. Then the weapon apparently also 'escaped' into several population centers in North America and the What To Think Network started blaming ticks. I think this corresponds loosely with another experimental, biological weapon assault on the population that was blamed on animals, this time in U.K. — 'mad cow disease.'

No viable excuse was ever given for either fibromyalgia or chronic fatigue syndrome, by the What To Think Network, of course. Before chemtrails made countless millions of people chronically ill serial killers had just been telling the sufferers of these two sicknesses that they were imagining their pain and ennui.

Apparently, the What To Think Network's science experts assumed that these ticks suddenly migrated throughout North America, especially into urban and northern areas, where ticks are never seen. I think it's obvious to rational people, by now, (numbering in the hundreds or thousands?), that Lyme disease is no more caused by ticks, than AIDS is caused by blue or magenta monkeys, or EBOLA came out of an obscure cave in the Heart of Darkness.

Fibromyalgia used to be a symptom of advanced trichinosis, and that's genuinely rare. What the serial killers are calling fibromyalgia is an endemic now, though, and it's been widespread since 1999. Extensive pain throughout the muscles and joints is a symptom that develops when one's immune system has been thoroughly compromised by bio-weapons, which specifically target the vital organs, over a period of time, through repeated exposure. The biologically engineered pathogens aren't hardy or particularly aggressive; they simply excrete a prodigious amount of acidic material into the bloodstream, while they're alive. All of the new chronic illnesses are from biological weaponry, of course, and it's been documented that these weapons are routinely deployed in the atmosphere and food supply.

Drinking water has been laced for decades with chemical weaponry; the Kalifornia Reich mandated that fluorides are even required in bottled, steam distilled water so if you ever go to Kalifornia, don't drink the water! Maybe extreme fluoridation is the main factor in why the majority of Kalifornians, since

World War Two, have generally been so vacuous, self-serving, duplicitous and non-committal, also perhaps why so many people who move there from better places, become the same way.

'Hepatitis C' is another new endemic that resists being cured by ordinary zappers since 1999. Before, the 'real' hepatitis C was cured in a day with zapping, because viruses in the liver are particularly susceptible to being quickly wiped out by micro-current for some reason. It was kind of fun to watch that fast transformation back to health, in fact.

'Severe candidiasis,' is another misdiagnosed new chronic illness. Zappers get rid of the candida in due time, but not the symptoms on which the diagnosis was based, because that problem, too, is just a sign that vital organs, hence the immune system, aren't working adequately. I think this one is the most popular new misdiagnoses, by incompetent naturopaths, (the majority), of the new chronic illnesses.

Hypersensitivity to chemicals and to impurities in food, are also general symptoms of the new deficiency sicknesses. Most minor cases of these new illnesses generally get corrected in a day or so with zapping. Most of what people call 'allergies', are simply temporary food intolerances.

The beardenesque, Stephen King tales of super pathogens from military labs that spread quickly throughout the population and kill millions in a day are designed to cause you to feel hopeless and paranoid, in the face of 'science out of control!' of course. The occult/corporate enemies of humanity don't want you to recognize that the only way they could prepare the population for genocide, was by incessant, heavy, frequent exposure to non-lethal bio-weaponry in chemtrails, packaged food, institutional cafeterias such as school lunchrooms, and however else imaginative, resourceful, institutional operatives could get it done with the tacit cooperation of our corporate, alleged governments.

The most bizarre pathogens we've encountered are aerially administered, experimental nanobots that carry toxic material to specific places in the body—a sort of reverse clathrating process, if you will. Dooney was stricken with that, right after a low-flying four-engine plane dumped a load of white powder on her neighborhood, while she was working outside, and since that happened, a couple of weeks ago Dr Steve has been treating a lot of new, local cases of pneumonia in his clinic. He buys zappers from us, fortunately. Someone in the predator safari crew found out on the internet, while we were all working on Dooney, that the US Air Force had been conducting field studies with this new weaponry on the populations of Wyoming, Arkansas and a couple of other states, and that the technology originated in China. I remember that five years ago people on the Flathead Indian Reservation, just north of Steve and Dooney, had been dropping like flies from aerial weapons assaults. The number of sudden deaths from respiratory failure, on that Reservation, seemed to be far beyond the proportion of respiratory failure deaths in the general population. The sewer rats do like to experiment on people of color, as they openly indicated on the What To Think Network, a few years ago.

I think we messed that ad hoc Chinese/American corporate military hierarchy and agenda, up enough that nobody else will be troubled that way by nano-assassins, and Dooney overcame the effects of the assault and recovered within a few days.

Our psychics regularly go out of body to search the ethers for corporate mass murderers like these. We know that the world order won't give up it's cherished hope of genocide, just because this network has disabled their extensive chemtrail program and death tower agenda, as well as sanctioned a few top rats in the international occult/corporate/military hierarchies. It's obvious to anyone who will look, though, that the loss of leadership in this world order has steadily caused their minions to lose a lot of cohesion and focus in the four years that we've been doing this.

Not long after the chemtrail program went into high gear, in the beginning of 1999 throughout North America, Europe and industrialized Asia, millions of people became chronically ill and, as you may recall, the initial stage of these illnesses was acute and chronic respiratory infection, which filled the hospital emergency rooms, and induced the immediate deaths of millions of infants and old folks, in the first months. Anyone who was using a zapper in those days remained clear of the new chronic illnesses, because killing the new pathogens at that stage and for the following years, prevented the weaponry from excreting huge amounts of corrosive toxins into the body. These toxins were intended to disable the vital organs, in order to destroy immunity. The lungs are an insufficient barrier to aerosol bio-weaponry, so just about everyone at least got lung infections in those days—remember?

Even though none of this was ever mentioned by the What To Think Network many people still remember the time when everyone got sick in early 1999, which marked the sudden, full scale onslaught of chemtrails throughout most of three continents and the Antipodes (Australia and New Zealand). Not much that's genuinely newsworthy gets mentioned by the What To Think Network, in case you haven't noticed, unless it's bad, discouraging news, of course.

Many, by now, are convinced that the first couple thousand orgonite cloudbusters prevented, by the middle of 2002, the initiation of 'phase two', of the chemtrail agenda, which was to have been the kill shot: aerosol anthrax, smallpox and ebola. The What To Think Network would have been blamed the genocide on Muslim terrorists, in order to persuade the survivors to embrace martial law.

Unless the population was first 'prepared', by compromised immune systems over two or three years through constant exposure to mostly non-lethal biological weaponry, exposing us to the killer aerosol weaponry would have been relatively ineffective, because not enough people's immune systems would have been sufficiently disabled. Killing only a quarter of the population, for instance, instead of two thirds or three quarters, would probably only have enraged the survivors, who would immediately have thrown down the occult/corporate hierarchy's ill prepared puppet governments and minimal police forces. By the time the death towers had been erected all across the planet, the population was officially supposed to have been reduced to half a billion, (around ten percent), as I mentioned earlier. The sewer rats jumped the gun, and the debilitating and intensely ugly death-tower forests, are now a graphic reminder to more and more wakening people that the What To Think Network's emperor has no clothes.

The astronomically expensive chemtrail and death tower agendas, just like World War Two, required many years of preparation before they were deployed. Note, for example, that nearly all of the aircraft, ships and heavy weaponry that were used throughout WWII, by both sides, were developed before 1937, and were mass produced in sufficient numbers by both sides, well before the outbreak of full-scale conflict. The war's beginning was merely incidental and inevitable, in fact. The military training facilities were all in place then, too, and in the case of Pearl Harbor, the Philippines, Singapore and other 'allied' fortifications in 1941, only the obsolete aircraft, equipment and ships were left as easy targets for the Japanese forces, a large fleet of which Roosevelt well knew were near, and steaming toward Hawaii—even the What To

Think Network tells Roosevelt's betrayal, though their pundits excuse him, rather than call him a traitor.

The good stuff was all deployed out of Tojo's reach, ready for action. That was the beginning of the end of Japan's brief, London-managed hegemony in Asia. Spain and China were the arranged proving grounds, for all that new, destructive technology, and the populations of those countries were the unwitting test-bed. Hitler's end was assured when Stalin sucker punched him on cue, early on.

It's well documented that Eleanor Roosevelt's darlings, the young American communist volunteers of the Lincoln Brigade in Spain, became the US Navy's OSS during WWII and then, conveniently, the 'anti-communist' CIA immediately after that. If Senator McCarthy hadn't become addicted to heroin, he might have achieved some success in routing the commies out of the US Government, but I don't think there would have been enough people left in Washington, DC, by then, to do much except turn out the lights, as they were leaving.

It was Eustace Mullins who first discovered, much later, that the root of the Cold War dilemma wasn't communism; it was, (is), the occult/corporate world order, and he coined the term, 'the hegemony of parasitism.' Communism and National-Socialism were just the hand-puppets of their global, cynical and gratuitously violent Punch and Judy comedy and National-Socialism 'won' in this case.

Chemtrails started showing up experimentally in the mid-1980s, and within fifteen years the technology was apparently ready for widespread, full time deployment. The only new development since 2002, was that many or most passenger jets are now equipped with chemtrail spewing nozzles and small holding tanks, so that they can spray the population centers at lower altitudes, during takeoffs and landing approaches.

Eric Carlson documented that on Etheric Warriors, last year. We believe this was hastily developed, after it became clear that the widespread distribution of orgonite cloudbusters had made the regular chemtrails, which are sprayed at between 12,000 and 20,000 feet altitude, ineffective, long before the spewed weaponry reached the ground.

What can't count as a new development, but is a significant modification, is that after people on the internet began describing the features of the white, unmarked chemtrail jets in 2001, many of the jets started sporting red tail sections. It's easy to see these jets with binoculars, because they fly at much lower altitudes than the jets, which are making contrails. The jets which make contrails fly too high to be clearly seen with binoculars. Try using binoculars to compare chemtrail jets with higher altitude jets if you want to experience a distinct surprise.

I believe you can still see clearly through binoculars, as Carol and I did a couple of times, the outdoor assembly line where scores of retired, passenger jets from American and British airlines are gradually changed into white, windowless chemtrail jets. That's happening at Mojave Airport, which is near ultra-secret Edwards Air Force Base in Southern California. You'll need binoculars to get a good view, and you'll probably also see US Air Force refueling tanker jets on the tarmac, where the finished chemtrail jets are parked and ready to go. On any day you can count forty or fifty of these shiny, white unmarked jets, on the assembly line.

Evergreen Air Corporation's restricted base north of Tucson, adjacent to Interstate Highway Ten, has been a staging area for chemtrail jets since the beginning. You can see dozens of them on the tarmac on any given day, as you drive along the freeway between Tucson and Phoenix. Evergreen Air is a CIA asset, and they're the corporation who also sprayed the entire population of Southern California with chemical weaponry in the 1980s, using thousands of 'civilian' helicopters, in low passes, in that case.

I suppose anyone who wants to log plenty of hours of multi-engine flying, in hope of landing a good-paying job with an airline, could fly these jets, though it may be that Russian, Chinese, North Korean and other East Bloc military pilots are flying them all. I'm not aware that any of the pilots of these countless thousands of omnipresent, unmarked jets have ever spoken to anyone publicly, or even privately, about their work. That level of secrecy parallels that of the technology of HAARP and the new death towers, and implies military security and secrecy oaths; perhaps a language barrier, too.

If the PJ folks were any sleepier, the corporation would probably just make passes over all the population centers at 2,000 feet or so, as happened the first time I got sick from chemtrails in San Bernardino, California in November, 1995. In that case I was mystified to see an ancient, four engine Air Force cargo plane laying spew in lazy circles right overhead, just under cloud level. I was sick before the end of the day, and I remained sick until the first time I used a zapper, five months later. I've never been sick since that day, thankfully, except for a couple of days in January 2003, not long after I was injected in the left forearm with a bizarre poison, and six of us were marked repeatedly on our chests with poisoned needles. D Bradley still has the 'L'-shaped scars on his chest, from that nocturnal visit, and I have a strange scar on my forearm from the injection. Battle wound scars! Carol told me that the letter, 'L' was marked on our chests, repeatedly in italic form, because Bradley and I had been exposing the cynical, Theosophy agenda ('Love and Light').

The spraying didn't let up until long after it was no longer effective, but the physical characteristics of the chemtrails had changed, by the time orgonite cloudbuster distribution reached critical mass, in mid-2002. Before, when two or three white jets started making tic tac toe patterns in the skies over cities, the spew quickly spread out, destroyed the cumulus clouds and whited out the entire sky, (a process of static colloidal dispersal, enabled by deadly orgone radiation, which the weaponry produces), as it saturated the atmosphere, on the way to the ground and into people's lungs. After a cloudbuster was set up in an area, the chemtrails no longer completely spread out, and they no longer interfered with cumulus clouds.

In the orgonite cloudbusters' earliest days—the six month period between the beginning of the expansion of this network and the sudden, worldwide erection of the death tower network—one could see characteristic, huge blue holes in the sky, muck over areas where there was a cloudbuster and, in the holes, some chemtrails disappeared instantly while others lingered for a few minutes and up to an hour. At the end of each afternoon, when the spew-jets all went away, (north, toward Canada in our case), the spew gradually dissipated, usually leaving the local sky in pristine condition. After the death towers were thrown up, in the fall of 2001, the cloudbusters' blue holes were suddenly much smaller but the chemtrails still disappeared at the same rate. It was pretty discouraging to witness our huge, lovely blue holes shrinking, but this movement has been characterized by a 'two steps forward, one step back', pattern of progress, as the other side ceaselessly invents ways to partially counteract the healing effects of orgonite in the atmosphere.

We believe that the death tower network's sudden erection was supposed to have been timed with the subsequent release of deadly toxins from the chemtrail jets, but 'phase two' of the chemtrail program was either never tried or was begun but failed, early on, due to the cloudbusters. It may be that the spew-planes overhead, whose spew disappears instantly, are the ones laying down anthrax, ebola and smallpox, though. As a rule, orgonite seems to absorb the worst concentrations of deadly orgone radiation, quicker than it absorbs lesser bad energy fields. Take away poison's energy component, and it's not a poison any more—it's oxidized or otherwise neutralized ionically. Most chemtrails these days are probably spewed out for psychological effect; merely deployed as disheartening graffiti in the sky. Have you noticed that the PJ folks, who are just now waking up to chemtrails, are crestfallen, whenever they see even a disappearing chemtrail? They're more upset, over chemtrail smidgeons, than the far fewer people who noticed them in 1999 were upset, when they really were a problem and looked, (and were), a lot more menacing. Human nature is a study in contradictions, but maybe that's what makes us all so interesting.

After the middle of 2002, fewer and fewer new cases of chronic illness were reported by our potential zapper customers. Toward the end of that bizarre period, the most frequent complaint I heard from enquirers was 'severe candida.' I think that the term was simply in fashion with incompetent naturopaths at the time, as a preferred misdiagnosis. Most of our customers come to us after having done a little exploring among alternative practitioners, and after finally cutting their connections to the serial killers. Dependence on serial killers is a self-perpetuating form of willful stupidity, in which the victim is unwilling to take responsibility for his own health because they believe, in spite of overwhelming evidence to the contrary, that MDs are actually competent and truthful.

I think the serial killers had prudently abandoned their monotonous, 'It's all in your head—stop whining!' diagnosis, after the number of chronically ill people suddenly skyrocketed in 1999, because even their Pajama People cash cows knew that something was up, by then, and enough of the clinic pigeons started demanding answers, finally.

Before the drop in complaints of new chronic sickness in the middle of 2002, those reports were coming from about half the people who wanted our zappers. Before 1999, when the chemtrails became omnipresent, these reported problems made up less than ten percent. In Africa these problems are almost non-existent—chemtrails and pharmaceutical poisons are rarely seen there, and most of the bio-weaponry eggs seem to be in the World Health Organization's AIDS basket, on that continent. There are as many death towers, per capita, in Africa as there are elsewhere, though, and in most places, the underground power sources aren't' in place, so the death transmitters have to be powered by those enormous diesel generators in the fortified concrete shacks, at their bases. Each of these generators could supply power to an entire small town and maybe, after the fall of this world order, that's how they'll be used.

The serial killers, (MDs), seem to be universally obsessed with misdiagnosing these illnesses, also with prescribing antibiotics, which always make these problems immediately worse. MDs consider these millions of sleepy unfortunates to be their personal cash cows. You and I can't comprehend how someone could live by exploiting innocents, unless we were predators, of course, any more than we can comprehend why immaculately-clad, big-hair televangelists don't feel any qualms about buggering little boys, during their Sunday programs' commercial breaks.

Antibiotics are extremely potent, acidic mold toxins—that's right!—the same stuff that fungi excrete into your blood, lymph and cerebrospinal fluid, in order to weaken the surrounding tissue and turn you into walking mushroom farm or stinky-cheese factory. The reason antibiotics, destroy bacteria is that even bacteria can't survive the effects of the more powerful fungal toxins. Other fungi thrive in that environment, of course, so when you get suckered or bullied into ingesting antibiotics the poisonous fungi in your body are simply doing what the Bible advises, 'Be fruitful and multiply!' I'm sure you realize that 'antibiotic' means 'opposed to life.' That pretty much sums up all MDs' unspoken credo. I wonder how many of them cough, stammer and avoid eye contact, when they read the Hippocratic oath aloud.

If you were a fungus, and had to contend for sustenance in a human body with a bacterium, wouldn't you, too, want to eliminate your competition, by any means possible? Anaerobic fungi thrive in a much lower pH, (more acidic), environment than anaerobic bacteria can, which is why the world order favors deploying bio-weapons, like mycoplasma and penicillin, which are fungus-derived. Maybe God will turn them all into toadstools, and life will suddenly get better for the rest of us. Aerobic bacteria in the intestines promote health, and they enjoy micro-current. Antibiotics destroy these beneficial bacteria and poisonous bacteria and fungi take their place. If you want to see how zapping helps the good little guys, notice that zapping around the clock turns your feces a very light tan, (the color of healthy, uninfected bile) and feces from an infested intestinal tract is very dark, due to the overwhelming presence of poisonous, acid-producing, anaerobic bacteria. Good, useful, scientific experiments aren't restricted to institutional labs, of course. 'Ask Mr. Wizard!'

An added bonus for serial killers, is that after one of their PJ pigeons has taken a course of antibiotics his colon wall has become permeable, which means that worms, bacteria, fungi and viruses that had been confined to the intestinal tract can then travel throughout the body and choose new homesteads that are more to their liking. This generates an entire range of new serious illnesses, for the serial killers to incompetently 'practice' on, many of which are fatal, but all of which are ultimately debilitating to the victims, who then become perpetual fonts of wealth for the serial killers—well, not perpetual, actually, since they're being killed and financially bled on the installment plan.

Parasites in the brain cause depression and apparently boost schizophrenia. Did you stop to think how many scores of millions of Americans are on prescribed brain drugs, to control their depression and behavior? How many drug-pushing shrinks would be out of work, if everyone just used a zapper to cure depression and feel balanced? Brain parasites often interfere with the endocrine system, too, destroying the balance of one's metabolism and emotions.

Parasites in the heart, including big flatworms, account for most heart illness; parasites in the pancreas cause most cases of diabetes and low blood sugar; parasites in the liver cause hepatitis and other problems; parasites in the joints cause arthritis; fungi eat holes in the arteries, which the body wisely plugs with cholesterol, which causes atherosclerosis, high blood pressure, strokes, heart attacks, etc. Those are the high points; most health problems are directly caused by parasites and most of these parasites' early ancestors in any human host, came out of the colon, after a serial killer prescribed antibiotics.

I'm told that the old sulfur-based drugs actually worked, and that's why they were eliminated by the pharmaceutical cartel, during WWII. Similarly, the SS men who run that cartel out of Switzerland, banned colloidal silver in the 1930s, though countless thousands of people are making their own by now.

I know the same cabal desperately wants to ban zappers, but that would be like banning toothbrushes, since anyone can make zappers, and they don't want to draw attention to something that is going to expose them and get them all arrested, sooner than later. For that matter, colloidal silver started to be sold openly in the 1990s, in spite of the legislation that was passed that guarantees punishment for doing that. It's well known that anyone can make superior CS in a few hours with a nine-volt battery, a couple of silver wires and a cup of distilled water.

Even PJ folks recognized the benefit of colloidal silver, so the drug cabal didn't dare mount a major legal campaign to stop its dissemination; they had to content themselves with disinformation—sniping by paid sociopaths in public forums, instead, which failed. This is mainly how they've tried to slow down the dissemination of zappers, too. Carol and I are now only getting persecuted by the Canadian postal system, which delays our airmailed zapper shipments to distributors, in that country for over a month, unless we pay extra for express mailing, in which case the packages are only delayed a week. The fake-zapper scam, that targeted our business and reputations last summer came out of Canada, too.

It's worth mentioning that in ten years of making and selling zappers, I've never been able to send one to a California address successfully, unless delivery confirmation or a certified mail receipt is attached, or I've sent it by express mail. When we occasionally forget to attach a tracking document, the package simply disappears. It's pretty weird, don't you agree? California is the most enthusiastically fascist state in the US, though, so this shouldn't be too surprising.

After Reich was railroaded into prison for curing cancer and, I think, right before the feds conducted their public book burning, the US Congress passed fake national laws banning the manufacturing, distribution and even personal use of orgone-related technology. Many thousands of us are 'breaking federal laws' by making, using and in many cases selling our orgonite devices, so why do you suppose the American Gestapo hasn't tried to shut us all down, even though the What To Think Network's national newsreaders have called orgonite, 'harmless terrorist devices?' Wouldn't that make us 'harmless terrorists' if those newsreaders were telling the truth?

Serial killers, at least in America, won't say the word, 'parasite.' If you want a little fun, and are still ignorant enough to consult those street-level drug dealers when you're sick, bring up the subject of parasites and watch them squirm, like parasites that have been dragged out into the light of day. To be fair, one of my zapper customers lives in Romania, and her MD had told her that she has a specific parasite, (though he was clueless about how to get rid of the infestation), so not all MDs are terrorized by the word, 'parasite,' apparently.

I know that our Ugandan cohorts and friends, Rushidie Kayiwa and Paul Batiibwe, who are MDs, are not serial killers, because they prefer to use alternative medical approaches rather than pharmaceuticals. Dirk Verelst, an MD and psychiatrist, (and gifter), in Amsterdam loathes pharmaceuticals, and promotes natural healing. He keeps orgonite devices, including one of Cesco's beautiful and practical, 'Little Secrets,' in his office, for patients to handle, too.

Once in awhile the global drug cartel induces the What To Think Network in the UK, Canada and the US to initiate half-hearted, disinformation campaigns against zappers, in order to test the waters, but it always has the opposite effect and increases zapper sales, so they quickly abandon the ploy. I think the harassment we're getting, through the Canadian postal service, is just sour grapes, on the corporate/government end. The US Alleged Government is the only one that imprisoned a zapper maker and that was a couple of years ago. I don't think that fellow was a fighter, though, because the things he was imprisoned for were pretty innocuous to the Feds, compared to my own public profile and activities. There are about a dozen zapper makers in the US now, but when he went to jail, I think he and I were the only ones making and selling zappers in any quantity, in the US. Dr Clark had apparently moved to Switzerland when the $#!+ hit the fan, right before that. You can see that the present zapper makers in the US, most of whom are our co-workers in the gifting network, aren't intimidated by this treasonous regime, any more than most of the makers and sellers of colloidal silver were.

By now, even the sleepiest PJ person distrusts the medical establishment, at least a little bit, and all rational people avoid the serial killers and their costly poisons. The prairie fire of rising awareness, which zappers and other effective alternatives are fueling, is building momentum and will bring a decisive end to the tyranny of the dope cartel and their impoverishing, oppressive hospital Gulag Archipelago before long. An awful lot of MDs and other dope cartel operatives will be held accountable for mass murder, or will have to go on the lam and turn to more straightforward, confidence scams or panhandling. Yet another schizophrenic aspect of the West, after all, is the way countless thousands of grossly incompetent MDs are treated like royalty, and can prosper even though it's common knowledge that they're doing no good at all and, in fact, harm just about everyone they touch. Their position is pretty precarious, under the circumstances, and I think most of them know that by now.

One rung, literally, below these serial killers are the vast, new cadre of incompetent but generally harmless naturopaths, homeopaths, acupuncturists, 'body-workers,' ad nauseum who make up the cynical, worthless net that catches most of the people who become dissatisfied with their 'family practice' serial killers, and seek 'alternative medicine.' Health food stores are chock full of little bottles of pills that are aggressively promoted, (by inference), as cures for specific illnesses, but which never seemed to work for you and I.

Schools like Bastyr College of Naturopathy are set up to train people to only 'treat' illnesses with natural means, but not to cure them. A graduate of that school once told me, by the way, that they get more real training in omni-sexual orgies than in medicine there. Dr Reich pointed out that sexual promiscuity is a sign of emotional plague.

Most chiropractors are cynically trained to only realign the upper spine, leaving the foundational imbalances in place, so that the upper spine will soon be knocked askew again, causing pain and resulting in more money for the chiropractors. At least they don't eventually murder their cash cows, though, as MDs mostly do. As con artists know, 'You can sheer a sheep many times but you can only skin her once.'

These Machiavellian medicos' tactics shouldn't surprise you. If you were in the same precarious position as the deceptive agents of the pharmaceutical cabal, had no character or conscience, but controlled nearly unlimited human, material and What To Think Network resources, it would be in your best interest to also corner the 'alternative medicine' market, just to be safe.

The petroleum cabal bought all the patents on alternative power-generating technology, eliminating all competition in the process and preventing the further refinement of those alternative energy devices. Strident, Rockefeller-funded environmentalists promote these crummy, but costly wares as, 'The alternative to polluting Gaia's atmosphere with petroleum products.' Whenever the cartel had the opportunity, they also bought viable free energy device patents and ideas, which is what happened to Viktor Schauberger's wonderful inventions in 1953, also the efficient, water-burning internal combustion engine, in 1914. Otherwise, the Corporation still occasionally has a higher-profile free energy device inventor murdered to scare the rest into remaining silent, which is what happened to my friend, Wilhelm Muller, a couple of years ago and to his friend, Brian DiPalma, a few years before that. The oil companies sell all of their inexpensively produced and underwhelmingly effective solar panels and wind-driven generators dearly—pricing it all out of the market, for the third-world people who would actually benefit from it all, otherwise.

When my first son was born, he had such severe colic, that we couldn't get any sleep, so X-1 and I hunted for a naturopath who could find the source of the problem. That was in 1979, but we already figured out through experience, by then, that most naturopaths are incompetent. We knew that asking an MD to help, guaranteed that the problem would just be made worse or might even get the baby handicapped or killed.

At the time, one was lucky to find even one competent, old school naturopath in an entire state or region, and those were usually under the AMA's gun and elderly. The incompetent ones had no trouble at all staying in business, and there were only a few of them, back then. By now, you can't p!$$ behind a bush without hitting an 'alternative healer,' but the upside of this is that their grand-scale numbers are a clear signal, at least, that the market is finally demanding an alternative to serial killers.

We were eventually directed by a friend, to Dr Shirley Snow, in Derry, New Hampshire, who figured out that our baby was simply lactose intolerant, and that the problem was coming through the mother's milk. X-1 quit ingesting dairy products, and the colic stopped abruptly. What do you suppose a serial killer would have done about that problem? He'd 'practice' on the baby, subscribing antibiotics and a grocery list of other, progressively more toxic poisons that might have debilitated or eventually killed the kid, of course. This is the sort of treatment that National-Socialism has made 'free', in places like Canada, and has successfully scammed countless millions of wage slaves in America to numbly suck cheap medical insurance and brutal, incompetent, HMO coverage from the corporate teat. Yummy!

When we were on our way to Tonga in 1984, after I finished a job in Western Samoa, we stayed in American Samoa for a couple of weeks so that I could learn to operate an offset press in preparation for a publishing job in Fiji. We met a couple in American Samoa who also knew Dr Snow, and here is their story:

Several years before we met them, they had returned to Boston from living in Haiti, after their two small children had developed leukemia there. The children were absorbed into New England Medical Center's oncology mill, and one of them soon died from their poisons, while the serial killers were 'practicing' on them both. The parents abducted the other child and took him to Dr Snow, in New Hampshire, who soon cured his leukemia. The serial killers in Boston quickly had the parents prosecuted by the Commonwealth of Massachusetts, for taking their boy away from them and arrest warrants were issued, so they couldn't go home to Boston. This is commonly done in America, to make examples of people who take responsibility for their own children's health. They soon moved with their healthy boy to American Samoa, instead. That was long before the treasonous US Congress federalized all victimless crimes, fortunately.

The woman who had first told us about Dr Snow had developed tuberculosis of the bone, when she was living in India and had to return to the US. Having run the gauntlet of serial killers, including some at Walter Reed Hospital in Washington, DC, who 'practiced' on her for no charge, just to satisfy their curiosity, and also misdiagnosed her sickness, she somehow found Dr Snow soon after that, who correctly diagnosed the problem, and she was healthy again, within a month or so. All of that was in the day before zappers, of course. We knew her before she went to Dr Snow, so we personally witnessed her fast transformation to vibrant health.

The American Medical Association have routinely been murdering and imprisoning healers in the US, since the Federal Reserve Corporation, (the City of London), was given control of the US Government, in the 1930s by FDR and his criminal Congress creatures. The AMA went after Dr Snow so many times, that she probably stopped keeping track, by the time when we met her. The reason they can't touch her is that New Hampshire folks have a more progressive attitude about personal freedom, than the rest of North America does, and she's had thousands of happy customers in New England and beyond, many of whom, (like me), are quite outspoken.

I'd be astonished to know if there are more than a dozen competent, courageous naturopaths like Dr Snow and Dr von Peters, left in North America. The 'genuine healers' trade is a small community, though, so we do come into contact with good doctors. The only MDs I personally know, though, who are in that category are Rushidie Kayiwa, Paul Batiibwe and Yahya Sekagye in Uganda, who primarily cure and heal sick people with potent natural remedies and zappers. I have a lot more to say about them, especially Doc Kayiwa, who is distributing orgonite throughout East Africa and is thus reducing the ongoing threat of CIA- and British-sponsored terrorism in that region without shooting anyone.

Dr Steven Smith in Montana was trained as a chiropractor, and also apprenticed to several gifted and accomplished natural physicians and healers. He has a wonderful track record, curing and healing people through the use of chiropractic, nutrition and energy medicine. I can tell you from experience that after he aligns one's bones they stay in place from the ground up. He's doing a lot more than just manipulating the joints, though—the man is a powerful, hands-on healer, who is adept at finding potent nutritional solutions for imbalances. I'm sure there are more in his league, but he's also in the gifting network and we lived a few hours from him for a couple of years, so we know him personally.

Dr Ed Group in Houston has www.ghchealth.com and buys a lot of zappers from us. He, and a crew of other chiropractors with social consciences, disabled most of the death towers and HAARP arrays, (several thousand towers), in and around that city, a couple of years ago. He was targeted by the AMA assassins, for curing a lot of cancer victims, some years ago, and decided it would be prudent to rather sell good healing products to a wider customer base, on the internet, so they could heal themselves.

Dr Rudi Verspoor has a renowned and progressive college for homeopaths in Ottawa, Ontario (www.homeopathy.com is his site). He combines the teachings of Dr Reich with Dr Hahnemann's, into a new, synergistic approach. When I visited him a couple of years ago he told me that curing illness is an act of aggression because diseases have their own consciousness and will. He teaches that healing is rather a loving activity. These principles also apply to interpersonal and even global-scale curing and healing pursuits, of course. Dr Rudi and his wife, Patti, (who is also an accomplished homeopath), have a cloudbuster, and disable the death towers in their region. Dr von Peters never met Dr Rudi, but he told me that Verspoor is the most accomplished homeopath in North America. Genuine pioneers aren't overly proprietary.

The principles of curing and healing apply to the body politic of humanity—why would it be otherwise? I've been trying my best, over the years, to convince people that along with healing our world, we have to deal aggressively with the disease that afflicts it, which is tyranny. Most people who are nice lack intelligence (another form of aggression is curiosity); most intelligent people just aren't nice, and they add to the world's burden. Our challenge has been to find other folks to work with, who are both sweet and intelligent and I have faith that there are enough of us to turn this world around now.

If you think aggression is inappropriate for fixing this world, consider this:

A Saville Row suited mega-parasite named Lord Nathan Rothschild isn't going to stop exploiting tens of thousands of enslaved Bolivian Indians in his vast Bolivian strip mines, if you gently hug him and tell him that you love and forgive him. He has to be dealt with in an appropriately aggressive, (lawful), way, if we want his corporation to stop murdering and exploiting those people. If you're an attractive, young man or boy who rather feels that loving Nat enough would end those people's torment, Nat might return your affection, but maybe not in the way you expected, and I doubt he'd close down Rio Tinto Zinc, on your account. Sanctioning a global-scale criminal like Nat would be part of the cure, for the social/economic/political illness he and his close associates are causing. Giving those Bolivians an opportunity to genuinely profit from their labor and to improve their lot through education, training and appropriate economic help, taken from imprisoned Nat's bottomless pockets, perhaps, is rather a healing, loving activity.

I think the silly assumption that 'love fixes everything in this world' is one of the fruits of dualism—tyrant-enabling, fake detachment, which the vast majority of sweet people buy into. The parasitic world order has always ruled only by default.

Now, enough of us are saying, 'Enough!' and are acting on our commitment to end their rule in appropriate ways.

If you believe that institutional predators have ever willingly given up any of the wealth and power they've pirated, then someone has successfully subverted your rational faculty. Even their charity has ulterior motives, in fact: Andrew Carnegie donated a small share of his billions of dollars to set up town libraries throughout the US in the 1920s, but that was the beginning of corporate censorship in America, since the conditions for accepting the money was to also accept only corporation-approved books for the shelves. The formal ascent of that London Corporation to the position of Sovereign in America in the 1930s was the end of a long process, not an isolated event.

Three and a half years ago, Dr von Peters was called before a federal hearing board in Atlanta, Georgia, to defend his livelihood. The consortium of naturopaths based at Bastyr, in Seattle and the Northwest College of Naturopathy in Portland, Oregon were trying their best to make it impossible for all naturopaths who didn't graduate from those two schools to have a license, and von Peters is the most prominent naturopath in Georgia and also represented a much higher standard of medical training, so that cabal of incompetents were trying to make an example of him, through government-assisted, public humiliation.

He told Carol and I about his dilemma, and we asked people in the network to help get the Doc out of the trap that had been set for him. He was ordered to show up as the defendant in that federal hearing, in Atlanta the following week, so Josh Clark volunteered to thoroughly gift Bastyr College, Seattle and the Northwest College in Portland, Oregon while Carol, myself and a few other folks tossed energy at the six or seven ringleaders and lawyers of the offensive naturopathy cabal, and whom the Doc had identified for us.

Carol made up an orgonite device especially for the Doc to take into the hearing room with him, which he received in time. Orgonite, which is half metal by volume, doesn't set off metal detectors, by the way. :grin:

The entire cabal were put off balance enough with Josh's gifting, our blasting and the orgonite in the hearing room, that they made monkeys of themselves, in front of the Feds and the assembled crowd. Doc, who gave lucid and inspiring testimony, came out of that hearing room smelling like a rose, and the backstabbing-naturopath cabal, (we refer to them as 'Bastyrds'), hasn't been able to touch him since then—that was in the fall of 2003. If those Bastyrds weren't predators, all the orgonite that Josh secreted around their campuses would have just made them happier, instead.

More recently, the Georgia Medical Board tried to remove Dr von P's license, so we tossed energy at each of them, in turn, and the case was dropped. Last month the Montana Chiropractic Board made a move to cause trouble for Dr Stevo, and we gave them the same treatment—problem solved and case dismissed. We've rarely missed, so far, praise grid.

Von Peters had bought a cloudbuster from Andy Schwarm of www.ctbusters.com, and has been systematically disabling all of the death towers in and around Chattanooga, where he lives. He reports the requisite, dramatic confirmations he's seeing in the atmosphere and also feels that the quality of the local ambience has greatly improved. On our drive across the country, before that, Chattanooga was the only city we traveled through where there was still smog and ugly ambience from the death towers. Other travelers have indicated that the last, remaining smog belt in the US, reaches down along the western side of the Appalachian Mountains, from Pennsylvania to Atlanta and Chattanooga, is in that belt. For some reason the Doc's cloudbuster wasn't making much of a dent in that, which indicates a particularly hard target. It got much better, apparently, since he also disabled a large number of death towers. It was nice to finally meet him, then.

When Carol and I were severely poisoned by federal agents with beryllium dust and lithium in December 2004, we got help from Dr von P, and within a short time every trace of those poisons was gone. That level of poisoning by specific heavy metals probably requires comprehensive and supervised intervention by a skilled doctor to correct, because specific, toxic metals throw all of the body's mineral ratios off balance.

Another 'front rank' member of the gifting network, outside of the US, who had been responsible for thoroughly gifting a huge metropolitan area got poisoned with lead in an apparent effort to knock him out of the saddle but he, too, was quickly healed by the Doc. Since then, many people in the network have overcome the effects of federally-administered poisonous metals with his timely, affordable help.

Carol and I traded fifty of our Terminators for a Pro Vitalizer Plus, and now when we get smacked by the other side's psychics, (mostly Chinese military predators these days), in reprisal after our gifting sorties and predator safaris, we simply heal the physical wounds quickly with this hand-held device, while we go after our assailants. Recovery time after those assaults is minutes, now, rather than days. Most of those wounds are simply torn muscle tissue, but until recently they've also been able to physically wound Carol's heart.

We just got the newly translated user manual, for the device from Dr von P, and it's an astonishing, comprehensive read.

Anyone can use this device productively and intelligently, for healing specific organs and tissues by simply reading the concise manual and following its explicit diagrams. Understanding the physics is another story, though we suspect there are subtle energy processes involved, as well as electronics. Tesla produced dramatic subtle energy effects, which he apparently assumed were only electrical. The Tesla coil, for instance, puts out more blue, vital orgone than anything that any of the inventors in this network have come up with can do. Carol and I use a Tesla coil to boost our radionics efforts and kinetic vortex-creating devices, in fact.

A spinner dolphin, as I mentioned, apparently provided the final energy shield around her heart on her last trip to Hawaii—well worth the price of the airfare, rental car and hotel, I'd say.

Professional, dissociative-programmed, psychic predators get at us through weak spots in our etheric fields, of course, which they can see remotely. The only way they're able to physically wound us with energy is by exploiting those weaknesses, usually from behind or from the sides. We, on the other hand, assault them and their masters from the front, etherically, and with potent force. They can't shield even from those of us who are non-psychic, because our direct connection to them is through the pain they're causing us in the moment. I think they're just too stupid to just let go, or else they're unable to let go once they've initiated an assault. Maybe they're like a moray eel or shark, whose teeth point obliquely to the rear and are unable to turn their prey loose, once they clamp their jaws down.

Before orgonite and zappers came along we believe these secret police and military psychics were routinely killing lots and lots of innocent people, this way. It's an ancient dirty-magic principle: remove enough of someone's life force and he expires, leaving no trace of the cause of death. The Chinese military psychics are particularly handy, by the way, as are the African voodoo practitioners. The Africans are the best at it, in fact, but they're also less likely to be enthusiastic supporters of the world order's Europoid and ancient Chinese hierarchies. All of the dirty dealers in the world, serve the occult/ corporate world order, in one capacity or another, consciously or unconsciously. Europoid psychic predators make up the vast bulk of the Monarch Program's and Tavistock's psychic shock troops and are de facto Theosophists, as a result of the popular but stultifying popular mind control protocols in the West. They're pretty insipid and non-focused, compared to the African and Chinese predators, though, whose training is based on more ancient, refined protocols.

I mentioned 'body workers' as part of the cadre of poseurs, who make up the drug cartel's 'alternative healing' safety net, so I want to make the point that I've known some reputable energy healers, who actually heal people by manipulating energy. The best one I've known is James Hughes in Ashland, Oregon, who took me under his wing for a couple of years, and helped me reconstruct my shattered ego and heal my wounded heart, so that I could eventually initiate this global movement. I also knew a very good one in Newburyport, Massachusetts who actually heals people and is rational, but I've forgotten her name. She started out with one of the standard, cultish fake healing schools in the area, but had a lot of undeveloped talent, so she got something worthwhile from the training, then went on to develop her own more viable, unique healing system. I know there are more like her around, who actually heal people.

James had been a successful Maytag dealer in that town and bought a new Mercedes each year. James is a guy who enjoys comfort and abundance, still, but when he was struck by a ball of etheric lightning in 1979, in his mid-forties, he was instantly introduced to a vastly expanded awareness of reality and knew that he needed to start a new career as a teacher and healer. If you want to get a glimpse of what happened to him, watch the movie, "PHENOMENON", which apparently isn't very exaggerated, except for the inference that anyone who's 'that wise and good' is going to die pretty soon, probably from brain cancer, 'because the world is too evil to hold him.' James heals himself and sometimes asks others to heal him. All of us are in constant need of healing, of course.

He resisted his destiny and felt overwhelmed by all of the strange new information he had gotten, and in a short time he lost his business and home. He was so debilitated by inner turmoil that he was bedridden. He told me that the only time he was able to get out of bed, for the following five years, was when he literally dragged himself to the bathroom. He had two teenage daughters at home, then, and his wonderful wife, Rose Mary, stuck with him and managed to keep the family together for the duration.

James and Rose Mary married in their teens, and all he brought to the union was a cardboard box full of personal effects. He had left school in his early teens, but he enjoyed earning money with his entrepreneurial skill, though. They prospered in a short time, and their prosperity increased until the 1979 event.

On the day that James committed to his destiny in 1984, his physical energy immediately returned and people started to seek him out for healing. By the time I met him in April 1998, he and Rose Mary had been traveling throughout the West in their big motor-home, which was loaded with crystals and other energy paraphernalia, conducting workshops to train people to properly use crystals, copper wire grids and other means to induce them to heal themselves and each other physically, emotionally and spiritually. Their Ashland home had become a sort of pilgrim destination for people from all over the world, who had heard of him and wanted to learn and to heal. Now, they only give workshops to groups of forty or so, and have stopped doing individual sessions.

James has been a positive example for me in other ways, too. For instance, he stresses that personal commitment is the most important thing in life, and that we mustn't ever be scammed into following gurus or joining cults. He stresses that in any given moment, we need to steer clear of judgment, belief and denial if we want to acquire new information/energy. He prefers the word, 'judgment' to 'prejudice,' perhaps because no matter how much we know, something new is going to come along to upset our personal paradigm, so there can't be any 'final decision' on how we are to perceive reality. Commitment and personal integrity allow us to assimilate new information and apply it to constructive ends, of course.

He never advertises or promotes himself, but when they pull into a new town in their motor-home people soon seek them out and they earn a good livelihood. I haven't mastered that, but thousands of new people each year find Carol and I on the internet, at least, and along with the free information and assistance they get, they usually also buy a zapper and Harmonic Protector. James started to prosper, after the initiation event in 1979, only after he committed to his new healing and education career. I'm sure he wouldn't want me to put his contact information in a book, but if what I've told you inspires you to look for him, you'll succeed.

James had no training, except for what he got in that instant in 1979, but that event also triggered his massive, latent psi ability and when he needs to know something, he pulls the information right out of the ether in an unadulterated form. You may have heard of people who can do that, but meeting and getting to know one is quite an educational and iconoclastic experience. It's likely that he's more easily able to do this, because he steers clear of belief, denial and judgment. A native American term for people who can do that, on account of having been struck by lightning is translated as 'lightning shaman.'

I think James represents the next level of human development, as is perhaps also characterized by the way dolphins relate to each other and to us. To get there, we need to cover some bases, and in our case that may involve educating ourselves about present, mundane problems, which we're all responsible for correcting now. Keeping our physical health, under the circumstances, is one of our daily challenges and might even be an obligation, if you consider that we're performing a needed public service.

I see a parallel between the way James has shown up in the world with his gift, seemingly before his time, and the way that Dr Reich showed up, seventy years ago, with basic scientific and spiritual information that perhaps is only now being put into practice, on a suitable scale.

You and I have no trouble grasping the concepts that Dr Reich introduced to the world, but even his well-meaning, supportive contemporaries, mostly failed to even glimpse this new science paradigm. It's a solid sign of progress that visionaries are no longer hounded down and destroyed, as happened to Dr Reich in the middle of the previous century, don't you agree?

James has mostly experienced acceptance, but the majority of people who attended the workshops I participated in were new-agers who squirmed, whenever he said the word, 'commitment,' so maybe he'll reach a broader, more sober and committed demographic, as the gifting network has, in recent years. He declined to be interviewed by Oprah, by the way, and I share his view that being courted by the What To Think Network is the kiss of death, for any genuine pioneer.

According to genuine physicians, finding the cause of any illness is the first step to curing it, and the skilled, experienced physicians who put their first efforts into finding the causes are sure to get the best results.

I appealed to Dr von P's research and development talent in the winter of 2003, when I felt that the market needed a single remedy or tonic to treat the new chronic illness endemics that were caused by the chemtrails. He was doubtful, at first, because he rather discourages sick people from seeking 'magic bullet' cures, but he began reading the available, documented research material about the contents of the chemtrails and recognized that the plethora of new illnesses were caused by a small number of factors, so he entertained the possibility that a single remedy might address the common, root causes of the new illnesses.

He spent the next six months gathering information from the internet and from books by reputable, research journalists like Dr Len Horowitz, before he considered formulating a cure. He knew that the weaponry in the chemtrails were aimed at compromising the immune system by specific and repeated assaults on everyone's kidneys, livers, spleens, and pancreas, so the remedy would need to include elements that could tone these organs. He also knew that including homeopathic elements, specific to the identified pathogens used in the chemtrails, would help reverse their damaging effects.

The body's immune response, which is the innate ability to repair itself and repel parasites, is based on those vital organs' ability to function adequately and in harmony with each other. Short of that, the body simply lacks the ability to repair itself or to clear toxic material away efficiently, which is why you often hear semi-conscious people constantly saying they need to 'detox.' Every competent physician knows that the main goal of treatment is to induce the body to regain internal balance and start healing itself again. The body is a lot smarter than wiser than any physician, when it is in balance.

Physicians who have integrity look for the causes of illness, then do what's necessary to get the body back into balance. Serial killers only aim to suppress symptoms, of course, which usually ends up causing more severe sickness later on. They count on that, in fact; it's their cash cow principle. MDs can't even suppress the symptoms of the new chronic sicknesses, of course; they only make these countless millions of new sufferers feel worse and worse, so MDs are goose-stepping on thinner and thinner ice these days.

You and I can't conceive how anyone can be like one of those cynical, predatory MDs, of course, because we're not parasites or predators. Really, in order to understand parasites, you would have to be a parasite but if you work in a hospital, cancer clinic, genetic lab or medical research facility you're no doubt seeing a lot of modern-day Josef Mengeles running around loose, with sharp knives.

When the Doc felt that he'd gathered enough information and pondered long enough, he came up with the formula for ChemBuster in a few minutes. When he had made up a large amount of the remedy, he sent me fifty of the one-ounce bottles and I distributed the new HomeoHerbal remedy to fifty of our customers, some of whom I knew and trusted, who were suffering from the new chronic illnesses and weren't getting decisively healed, with our Terminators. In most cases they were healed by the time they finished one bottle, taking twelve drops per day. In fewer cases it took two bottles to get it done, and in fewer still cases no significant progress was reported. The latter are the ones I strongly recommended to Dr von Peters for personal treatment. Those are the cases he prefers, anyway.

If one considers that only a small percentage of chronically sick people won't be able to heal themselves with zappers and off-the-shelf, effective remedies like ChemBuster, the implications are clear that physicians will need to acquire extraordinary abilities to find the causes of the illnesses suffered by this minority, and to prescribe appropriate treatments. I think there will always be a minority of people who are simply unable to cure their own illnesses by any means, so require competent professional help. In very rare cases, nothing can be done to cure a sickness; those few are the ones whom we reckon God wishes to remain sick in order, to learn a spiritual lesson. I think this is the real balance, not twisted, infantile ideological notions, such as original sin and past-lives karma, that were invented by nasty clergy, throughout the ages to keep the masses from empowering themselves.

This new, happy development implies the need for medical colleges, whose staff are qualified to train people who are already committed, insightful and have well-developed intellects, compassion and the gift of healing. Until now we've been offered an illusion of that workable paradigm by the pharmaceutical cartel's medical schools and the What To Think Network, but in days to come, humanity will actually require a much smaller, but genuinely trained and talented cadre of competent physicians like Dr von P who can cure and heal the hard cases for whom the popular, new, effective remedies are inadequate.

As it happens, Dr von Peters visited Uganda a couple of years ago to confer with Dr Rushidie Kayiwa and 'Secret Supporter,' and land has been purchased, northwest of Kampala, as the site for a new medical school. The university has already been chartered by the Ugandan government and it will focus on an integrated curriculum of naturopathy, homeopathy, herbal medicine, Chinese traditional medicine and allopathy—the world's first such medical school and the establishment of a new trend, I'm certain. I expect that this school will fill up with students from all over the world, once the funding is available to initiate it.

Coincidentally, the site for the new university is near the village where Kizira Ibrahim lives and not far from the district hospital Dr Paul Batiibwe administers, and in which he primarily uses alternative medicine and healing modalities with tremendous success, much to the expressed chagrin of the World Health Organization. Kizira is a master herbalist and psychic traditional healer, (also a fellow gifter and predator hunter), who has been Dr Paul's frequent consultant in recent years.

Dr Paul was the first African to make an orgonite cloudbuster and it was through my correspondence with him, which started in June 2003, that I eventually met Dr Kayiwa and 'Secret Supporter,' who hosted Dr von Peters and I on our separate visits. Doc Batiibwe is also an accomplished clinical researcher.

I visited Kizira several times, and on my last visit I made him an orgonite cloudbuster. I heard, a month later, that it had been bringing so much rain that the villagers were getting a little upset at him, so he took it indoors for a few days, and the rain stopped. Kizira and Dr Paul have initiated several gifting sorties to heal some of the traditional power spots in Central and Southern Uganda. Dr K took Kizira with him as a psychic consultant, on his first gifting sortie in war-torn northern Uganda. Fighting immediately ended there and the CIA-sponsored terrorists moved off into Sudan and Congo to regroup. Doc Kayiwa soon chased them further away by gifting those areas, too, and his efforts in Southern Sudan were so successful that he was invited to meet with the Arab government in Khartoum to discuss his work. We hope that this will parlay into further interest by the governments of Libya and Morocco. Doc K hopes to further exploit his fluency in Arabic to further the awareness of orgonite in those desert countries. We believe that the Sahara will become a garden before long, one way or another, from the intelligent, systematic distribution of orgonite, as Georg and crew are doing right now in the Kalahari.

Dr K and Kizira nearly emptied a couple of district hospitals in Northern Uganda, by distributing orgonite to the patients, who simply went home, because they suddenly felt well. The doctors in the hospital were mystified, but were eager to learn more about the uses of orgonite. Dr Kayiwa's recent visit to Khartoum was quite successful.

The sewer rat agencies simply can't conduct, even a well-funded and well-directed terror campaign from a distance, so the CIA and MI6 field agents who were training, equipping and directing those bloodthirsty terrorists in East Africa, found themselves in the same dilemma as 'all the king's horses and all the king's men' did after Humpty Dumpty had his great fall. Who would ever imagine that a few hundred dollars spent on orgonite and gasoline, (for the Doc's car), could have such a phenomenal result? Africa is the continent where orgonite seems to consistently get the most bang for the buck.

The City of London got pretty upset when General Museveni led a successful grassroots campaign from Tanzania to drive their last bloodthirsty, proxy dictator out of Uganda in 1986. London, in reprisal, immediately crashed the Ugandan currency then, forcing the Ugandan government to survive until now, by trading agricultural goods for the services of foreign governments, which turned into a fairly lucrative, mutually beneficial way to recreate the nation's minimal infrastructure.

A major highway was built by the Serbian government in exchange for food, for instance, and the Irish government built a couple of administrative buildings, also in exchange for food. Italy provided and maintains several medical labs, for a similar trade. I don't recall what the Ugandans wanted from Cuba, but a shipload of beans to that country was sabotaged by CIA operatives, who pumped seawater into the holds, en route, rotting the beans.

Dr Kayiwa and Georg Ritschl arranged to get some solar-powered stoplights set up, at several major intersections in Kampala, and that has eased the gridlock that often occurs in that city, on account of the complete lack of traffic signals. To the citizens' credit, the stoplights were warmly welcomed, and most everyone agrees to obey the signals, so enforcement isn't needed.

The CIA/MI6-sponsored terror campaigns in the north were London's further reprisal effort and a way to force the government in Kampala to divert money from needed infrastructure to national defense, but thanks largely to the distribution of orgonite, that threat has been eliminated, by now and the Ugandan Army can mostly stand down, at last. I mention London because the CIA has only ever been a subset of MI6, 'Her Majesty's Terror Service.'

Southern Sudan had been ravaged by another corporate-sponsored terrorist effort to drive the population into Uganda and Kenya, apparently because they wanted to get the uranium out of there without the inconvenience of having to pay for it; also, they aimed to debilitate the Ugandan government even more by forcing them to deal with a massive influx of refugees. That all backfired, too, of course.

On the Doc's first visit to Southern Sudan the people were sick, starving and dispirited and the area was in a severe drought. Animals were dying of thirst and starvation. On his return visit, a couple of months later, there had been plenty of rain, crops had finally been planted and were thriving, people were healthier and hopeful and the farm animals were robust. He induced all of that with a trunk full of towerbusters, distributed in just a few days. The return trip was when he got the invitation to meet the scientists and government leaders in Khartoum. He thought it was just a polite gesture, until the invitation came in the mail from Khartoum, right after his return to Kampala.

When he told the black Sudanese government people in the Southern region about our efforts to help Africa, they were surprised to hear that there are actually Americans who don't assume that Africans live in trees.

It helps that Dr K is quite fluent in Arabic, because that's Sudan's official language and half of the population, mainly in the north where Khartoum is, are Arabs. All of the Africans I've ever met are multilingual, by the way.

One of the refugees, Salva Kirr, who recently returned to Sudan, has gifted his ancestral village in Southern Sudan. He and his family had spent a few years as refugees in Kenya. His wife was offered a teaching job near Juba, the main city in the south, and we take that as a signal that the gifting network will expand there, too, in due time.

President Museveni gave farmland to the Sudanese refugees who chose to stay in Uganda, where there is plenty of uninhabited, fertile land and water. How many leaders do you know about in this world who consider 'more people' to be national assets, rather than a burden? Independent farmers have always been the backbone of progressive nations, of course, and people, everywhere, will mostly provide for themselves with honest labor, if given an opportunity. This is especially true in East Africa, where the people are mostly tribal. You may have been presented with an opposite view of Africans, but just consider the source, in that case: Africa's recent, colonialism-induced misery, has been fertile ground for data to be used out of context, by Europoid western democratic liberals, to pitch National-Socialism in their home countries or to exploit westerners' charitable urges.

President Museveni asked Dr Kayiwa for one of our Terminators, after Nathan Kagina praised them repeatedly on Ugandan National Television, Radio and in the newspapers. 'Secret Supporter' had originally asked Dr Kayiwa to give a Terminator to Nathan, who was at the point of death with AIDS then. Nathan is quite robust and energetic these days, as the 'before and after' pictures of him on http://www.worldwithoutparasites.com/testimonials.html show.

What Carol found out in September and October 2001, during her working visit to a village AIDS clinic in Western Kenya, is that ordinary zappers cure AIDS reliably. In some cases the recovering sufferers can use an extra boost, to get their bodies back in shape, of course. In light of that need, Dr Kayiwa is marketing an affordable herbal tincture, made from East African plants, to help them recover faster. Many traditional African healers are adept at combining herbs and ancient energy, medicine to help sick people recover their vitality.

Drs Kayiwa and Batiibwe took me along with them to a graduation ceremony and feast hosted by Dr Yahya Sekagye's 'students,' who are traditional Ugandan herbalists. The 'teacher/student' role gets a little fuzzy in that milieu, because Dr Sekagye seeks to acquire and catalogue their collective experience and knowledge and he teaches those village-level healers to prepare their remedies in a way that can be marketed in stores. The healers also take advantage of the gatherings, to network and to share information with each other. Getting together in a large group is something that just never happens much, for these village healers, because most of them are separated by distance and can rarely afford to travel. Marketing their specific remedies in stores is a fairly new, and potentially lucrative concept for them.

Dr Sekagye is Ugandan, but also spent a lot of time facilitating a similar effort in Cameroon, West Africa. He knows Credo Muttwa quite well, by the way.

All the time I was in that country, I wished Carol was with me, especially on that day when so many reputable witches were in one place. It would have been 'old home day' for her. She met some nice witches when she arrived in the village in Kenya, though they had only come to the village for part of the day, to perform a protection ritual for the American guests. It was Carol's first trip out of the US, and she encountered the most dire conditions imaginable, while staying focused on curing as many AIDS sufferers as possible. It was quite an initiation for her, in fact. The day before she left, she inadvertently let out that she could see and talk to departed people, and she was literally overwhelmed with tearful requests, by a growing crowd of people who wanted to know how their family members who died of AIDS were doing. She was crowded from the other direction by departed family members who wanted to reassure their loved ones that they were doing fine. Do you assume that being psychic is always fun?

I took my Zapchecker to Uganda, as I always do when I travel without Carol. She sees implants' energy signatures on people; the Zapchecker finds implants in people electronically. Most of the people we meet have injected and sometimes surgically implanted transponders and nobody can say what their precise purpose is, because all of that is a military secret, like the technology of the death towers, the chemtrail agenda and HAARP are. The only Ugandan associate of ours who wasn't implanted was Secret Supporter, and I suspect that's because he's not easily approached by bad guys, due to his political and tribal position.

Dr Sekagye, the consummate teacher and networker, had the most implants of anyone I tested there. He also had just contracted malaria, which he quickly cured with a little, basic zapper I gave him. Most of the implants are put in place with hypodermic syringes, which are disposable and are usually sold with medicine and implants already in them, for inoculating school kids or for shooting up the serial killers' pigeons in clinics and hospitals, in response to fake epidemics. They most likely are GPS-linked transponders, also perhaps used in conjunction with crystalline prions that are grown on nerve tissue, introduced through specifically infested wheat products. Zappers apparently destroy prions. The injected transponders are nanotech, just like the Pentium chip in your computer and much of the circuitry in your cell phone are. Nanotech weapons are discussed on disinformation shows like Rense and Coast-to-Coast AM, as 'future threats,' but they've obviously been deployed for decades against the population.

It only takes about a half hour for a strong magnet, like the one in our Terminator, to neutralize an average, injected implant. Any talented psychic can see their energy signatures, otherwise they can be found by anyone with a Zapchecker. Disabling them results in a pleasant, perpetual increase in physical energy and mental focus. People who have latent psi ability are able to develop that talent easier after the implants are disabled. Some of the larger, older ones have poison in them, though, which gets released into the blood when the implants are shut down without authorization, but if you get sick after disabling your implants, just drink a lot of water and, of course, use your zapper and the released poison will be neutralized, too.

Credo Muttwa recalls that in his youth, in the 1930s, children were being force-vaccinated en masse, by British-backed government medical agents, accompanied by soldiers. When the tribal elders figured out the scam, they conspired with the villagers to create similar blisters on their children's arms with heated corn kernels, so that when the government agents came to the village to force all the children to accept their molestations, the children could all 'show that they'd already been vaccinated.'

The elders were alarmed in the first place, because children who had been vaccinated suddenly lost their ability to see subtle energy and spacecraft. In those days mercury was apparently included in the vaccines to induce this effect. Mercury is still being included in schoolchildren's vaccines in the West, of course. Don't you wish we had some elders like those? The World Health Organization tries to enforce vaccination campaigns in Uganda, but most people there, including Ugandan doctors, just ignore them. President Museveni's not about to assign the army to enforce that predatory agenda, nor are the medical cadre interested in promoting it. Twenty-three years of London's proxy dictators in the Pearl of Africa left a sour taste in their mouths, for Europoid intervention.

If you want to get rid of your implants, order the cheapest model from www.zapchecker.com and when you've located yours, tape a neodymium magnet to each implant location. The neodymium magnets in our Terminators are ¼" thick and ½" in diameter and if you don't want to buy magnets but have a T, just hold the tail penny electrode on the spot for a half hour because the magnet is right beside that penny, inside the box.

A couple of years ago, the sewer rat agencies apparently developed a way to shoot these implants into acupuncture meridians with specially adapted cell phones and sunglasses from up to fifteen feet away. There are apparently special screens on the phones and the inner surface of the sunglasses' lenses that enable the agents to aim and fire accurately. Carol, Ryan McGinty and I were treated to graphic demonstrations of these new weapons on the day we went to Devil's Punchbowl with D Bradley to gift the human sacrifice site there, in advance of a major ritual. We got so many implants that day, on our way to the site, that we felt like pin cushions. They really didn't want us to put orgonite there, and each of us had quite a harrowing and unique experience then. All of that is in one of the reports in "The Adventures of Don and Carol Croft", available as a free download on www.ethericwarriors.com. That day was the only time I felt surrounded by predatory reptilians and draconians disguised, (not very well, actually), as humans.

I brought that up because after this new weapon was widely deployed, the zapchecker seemed to go a little crazy, lighting up along two feet or so of certain acupuncture meridians into which implants had been shot or injected. I discovered that if I taped a magnet anywhere along the meridian, I could then easily locate the implant. Carol and DB says they almost always aim for the heart meridian.

If you're out and about after having done significant damage to the predatory infrastructure and are feeling Eyes on you, just be especially aware of anyone around you who is fiddling with dark sunglasses, or is facing you and turning a cell phone in a peculiar way, and you'll probably avoid getting shot. None of this stuff has seriously harmed any of us, thankfully.

I consider Uganda to be the model for what might be accomplished, politically and perhaps economically, in any nation, where the City of London no longer holds the purse strings. The press is genuinely free there, and I felt entirely free of interference by psychic predators and pavement artists, most of the time, even in the city. There aren't many Europoids in Uganda, so it's hard for sewer rats to blend in.

Right after we disabled the two hundred or so death transmitters in and around Kampala, there was a rash of firings and resignations among the military and government elite. One or two of the popular newspapers openly criticize the president now and then, but none of the editors took exception to that comprehensive housecleaning effort.

Dr Kayiwa had also taken us to the few unhappy neighborhoods in and around the city where voodoo was being practiced and we gifted those places, too. The city is on several hills and there are very few paved roads, so I doubt we could have gotten much done if Secret Supporter hadn't kindly lent us his reliable Nissan SUV, for the duration. We got around to disabling most of the death towers in the more heavily populated areas in that country when I was there, and Drs Kayiwa and Batiibwe extended the effort far beyond that, afterward.

Orgonite gets astonishing, consistent and immediate results in Africa's atmosphere, for some reason. Carol and I have witnessed that again and again, and we feel that this is due to the relative vitality of the ambient orgone field in that continent.

Anyone can 'see' orgone. It's the same way you see the 3D imagery in those old posters that are made up of apparently random, overlaid dots of color. When you look at the posters the right way, that is, when you gaze at them in an alpha mental state, the image comes clear and seems three-dimensional. When you gaze at the sky or at an illuminated, plain surface in an alpha state, you see a constant display of moving 'squiggles,' both bright and dark. This is quite distinct, from the faint visual effect of blood moving through the tiny vessels in the retina. You know it's not the same, because the blood-flow causes the effect to pulse with your heartbeat and to move only along distinct pathways. Orgone is random and fills the visual field evenly.

Nobody can authoritatively tell what we're all seeing; maybe they're tiny, hyper-dimensional energy currents of orgone within the retina's cones and rods, and when they appear and disappear perhaps it indicates their movement in and out of our time/space reference, like the Sasquatch and dolphins are able to do.

I mentioned it because when Carol and I 'watched' the orgone field, while we were in Eastern and Southern Africa, the 'squiggles' were much longer, brighter, wider and enduring than, what you can see in North America or Europe, and the ratio of dark squiggles to light ones is much lower. Dr Reich apparently developed a simple scope that can be used to see these effects more easily, and this is one of the evidences that he showed Albert Einstein in the late 1930s. The translucent eyepiece of the orgonoscope might lend credence to the notion that one is seeing something inside the eyeball. Where better to see an omnipresent energy matrix?

When Judy Lubulwa, (this book's editor), first started making orgonite in Nairobi, which Carol told me was the most grievously polluted city she'd ever visited, the smog cleared away overhead, revealing vibrant blue sky, and it rained briefly every evening for many days afterward. In any un-gifted North American or European city it would take many times more orgonite, carefully placed at contiguous tower locations, to achieve that wonderful effect. There are as many new death towers, per capita, in Nairobi as there are in any other city in the world, and there's a lot of human misery and death there due to the AIDS epidemic, neo-colonial government corruption, predatory policemen and other London/UN-enabled horrors. Pretty soon, Judy and crew will have turned much of that around, of course.

Georg and Trevor's experience gifting Soweto and the incredible transformation that followed in that city is another good example of orgonite's potential in Africa.

James DeMeo, PhD, had visited Namibia, (Southwest Africa and the location of the Namib Desert, which is the driest region on the planet), to promote the use of the old-style cloudbusters in the early 1990s. The man who sponsored his campaign is 'Baron' Rostow, who has a home on the coast and owns a great deal of real estate, throughout the countryside and in Windhoek, the capital. 'The Baron' sought us out, when we were staying in Swakopmund, on the coast, and he wanted proof that the orgonite cloudbuster is a better option.

The morning he came to call, in December 2001, Karstin Rolloff, Carol and I had just set up a new cloudbuster in Karstin's courtyard, and it was poking a huge, persistent blue hole in the dense, morning fog bank that was rolling in, from the polluted South Atlantic. I don't know what sort of proof the Baron expected, but here was some dramatic visual evidence, at least. I think he'd heard that we came to Namibia to lend a hand to Gert Botha's effort in the Namib Desert. Over the previous months, Gert had blocked a score of successive, violent sandstorms, apparently including some especially severe HAARP-generated ones, with a hastily constructed, half-scale orgonite cloudbuster on his property, in the desert east of Swakopmund.

DeMeo left Namibia, ten years before, after none of his cloudbusters achieved desirable or even discernible positive results. I don't think there are many people who have been able to use the old cloudbusters effectively and safely. Dr Reich obviously had the skill, as does Trevor Constable, but we were pleased to discover, early on, that it's impossible to harm the atmosphere or ourselves with cloudbusters, even very large ones, which have orgonite bases.

An example of the old cloudbusters doing harm when not handled competently was when one of DeMeo's was left standing with the ends of the flexible tubes in a watering pond, on a cattle ranch in Namibia. The cattle wouldn't approach the pond and they would have died of thirst, if the farmer's wife, whom we later met, hadn't sensed the problem and removed the cloudbuster in time. After she removed it the cows were able to drink the water. The deadly orgone radiation in that pond was similar to the energy in the cooling ponds, over the underground nuclear reactors that power the new death tower network, by the way. Any psychic can see that dark energy.

If the Baron is prudent, he'll eventually decide to finance the construction and oversee the distribution of orgonite cloudbusters, throughout the Namib, the way Georg and crew have done successfully in the neighboring Kalahari to generate regular rainfall there. To be fair, the score of orgonite cloudbusters in the Kalahari were in place for about a year, and weren't generating rainfall until Georg later traveled to the Indian Ocean coast and disabled the scores of powerful HAARP arrays, between the Cape of Good Hope and Mozambique. Those transmitter arrays had generated an efficient moisture barrier that was moving the entire southern part of the continent toward famine. Look on a map and see how big that region is and how far the Kalahari is from the Indian Ocean coast, if you want to get an appreciation of what was accomplished by disabling those coastal HAARP arrays.

Carol and I swapped a new, full-scale orgonite cloudbuster for Gert's makeshift half-scale one, which we took a hundred miles north, through the desert, to the powerful vortex on Brandberg, the Namib's tallest mountain. On our way up a side of that isolated mountain, from the end of the road, we passed an ancient Koi San, (Namib bushmen), sacred site, which was within an enormous, pirated natural vortex. The site is a small, flat plateau on the side of the mountain, roughly fifty meters in diameter, which has several stone rings on it. Carol and I were hit with so much DOR, in that brief time, that it took us over an hour to hike less than a mile. As soon as we put the CB in place, the interference abruptly stopped.

That was a pretty weird day and that night; as we were camped by the side of the gravel highway, we were repeatedly buzzed by Vril flying saucers that sounded like a car driving over gravel. Every time that sound woke us up, during that moonless night, we raised our heads expecting to see headlights, but the sound came and went and we saw no vehicles on the road until mid-morning. We were too groggy to think to look up, instead. It's a pretty isolated area. On our way to Brandberg we had gifted around Swartkoppe, another isolated peak, which is reputed to host a very old, Vril underground base—a staging area, apparently, for the subsequent Vril occupation, under Neu Schwabenland, in Antarctica before and during WWII, according to what Carol was seeing. It didn't make a lot of sense to her, because she wasn't aware of the history, but when I told her what I'd read about that, she felt confirmed. Al Bielek had told us some related things about a failed American military attempt to rout the Germans from Antarctica in 1947, so it was good to get those confirmations in Namibia. The century-long Vril presence under Swartkoppe seems to be common knowledge in that region.

The second time we heard that sound was when we were

driving toward an enormous, bright orange UFO in Florida, a year before we were in Namibia. We had come upon a big, black triangle UFO that was hovering next to the highway, and whose crew were apparently focused on that distant, luminous, oval-shaped orange craft. We got out of the car, practically under the triangle, and saw halogen lights on the underside, so we knew it wasn't 'imported.' It banked steeply and flew slowly away from us. I think human, quasi-military UFO crews are kind of shy. I wish I'd had a cloudbuster to point at it, because I think I could have downed the thing. I bet that rude landing would never be reported, on the What To Think Network.

You might have seen one of these big, black triangle craft because, by now, they're apparently almost as common as the silver disc UFOs that have reputedly been mass produced on the Gold Coast, north of Vancouver, BC, by an American/British corporate consortium since 1947, according to a well-documented book I read, that was written by a very thorough, German research journalist.

Shortly after we set up house on the rural property near Spokane where we encountered the Sasquatch, I heard a sound that seemed like a C-130 transport plane was buzzing us at treetop level. Carol peeked out and told me it was one of those triangle craft. When I went outside the sound stopped, and I heard the characteristic car-on-gravel sound, instead, so apparently the real volume of the engines in those craft is subsonic—the house-trailer simply resonated to that low frequency and was vibrating loudly. That's the third time I heard the sound, but I wasn't able to see what Carol saw that time.

Carol and I have seen lots and lots of UFOs, since we've been together, and I reported most of that in my posted journal entries. Keep your eyes open, when you're out gifting, and you'll no doubt see some, too. Jesse Zaloudek, who gifted all of the mountaintop arrays and coastal HAARP arrays, in and around the San Francisco Bay area, photographed a large, dark saucer UFO, that was apparently absorbing DOR from the top of a really big death tower. I didn't save the photo he sent me, unfortunately. It was quite explicit.

You'd be surprised how many people, mainly observe the tops of their own shoes, and each others' derrières when they're out and about.

You've patiently read this far in the chapter about health and might be disappointed that I haven't mentioned food, but I've assumed that you already understand that it's better to eat fresh, organically grown food as much as possible, and to avoid consuming excessive amounts of white sugar, dairy products and meat, especially if any of that comes from corporate sources.

It's possible to acquire unbelievably radiant health by faithfully following raw, live food protocols and some rare individuals have been able to live well on nothing but air and water, but in practical terms that much health and awareness, effectively separates us from the rest of humanity. I learned that from direct experience, when I was young, lived on raw food and did a lot of fasting. It would be nice to get back to that sometime, but not now.

Otherwise, if you'll follow the simple advice I laid out in this chapter you'll most likely never suffer a debilitating illness, at least, and that's really saying something these days, when serial killers and the What To Think Network have turned most people into walking worm farms.

Well, this wraps up my first publishing attempt. Unless a lot of these books sell, I probably won't write another one, which might be too bad, because the best is probably yet to come for Carol and I and I love to share information and adventures. I don't have time to post reports as much as I was doing before because the zapper business—production and correspondence—is taking up more and more time as public awareness of zappers continues to expand.

Some voice communicators who are ugly do better on the radio. Communicators like me who aren't personally charismatic do better with writing than speaking. I'm always surprised when people tell me they feel inspired by my writing, because I get a sense that their eyes would glaze over if they heard the same words from my lips, instead. Before the internet came along, I don't think I had a genuinely candid and extensive conversation with more than a dozen souls, in my lifetime until then. Many people in this network, including Carol, have felt isolated, that way.

I had the time to do this book because Carol took a trip to Hawaii for three weeks in March, then went to Idaho for a week at the end of May. I started the book, tentatively, on her birthday, June 19, last year. That was intentional and I'm finishing the final draft on my birthday, May 5; that wasn't intended. I need to be alone when I write more than a few pages, but it's no fun being apart from Carol that long, I can tell you. I'll go pick her up at the Palm Beach Airport in a couple of hours.

Our expertly repaired and overhauled Zodiac Pro 650 was delivered from Miami this afternoon, after a two-month absence, which means that a pelagic adventure is probably imminent for my bride and me.

I honestly hope I didn't jerk you around too much or tax your credulity beyond its limit. The problem with considering the subject matter of this book is that, unless you're willing to adopt a broader view of reality than you acquired in college or from the What To Think Network, you're not going to develop enough forward momentum to have a whole lot of genuine, heart-stopping fun in this life, and I want you to enjoy yourself, while you're dismantling global tyranny's infrastructure and hierarchies and healing our world.

How much happiness can you stand?

ISBN 141209565-4

9 781412 095655

Made in the USA
Lexington, KY
07 December 2015